Innovation is thought to be essential in the context of public procurement for a variety of reasons, chief among them the potential for enhanced technologies, improved productivity, and greater inclusion. Still, more insight has been needed that connects scholarly work and practice. Dolores Kuchina-Musina and Benjamin McMartin make significant inroads in addressing this need – offering an exciting new text that is concise but still informative, inviting readers to consider procurement innovation concepts along with useful case examples.

—**Dr. Christopher L. Atkinson**, *Associate Professor, University of West Florida*

Public Procurement for Innovation

In this book, nationally recognized public procurement experts Dolores Kuchina-Musina and Benjamin McMartin present a comprehensive analysis of the alternative contract vehicles used to promote innovation in the United States (US). Kuchina-Musina and McMartin begin by introducing the innovation policy environment in the US, addressing current trends in the workforce, decreased investment in research and development (R&D), and how technology is increasing at a rapid speed. They then go on to discuss key terms and subjects to show how public procurement and innovation policy are tied together. Diving deeper, Kuchina-Musina and McMartin examine the pathways the federal government uses such as the Federal Acquisition Regulations (FAR) part 12 acquisition, a review of the Small Business Innovation Research (SBIR) and Small Business Technology Transfer (STTR) programs, and non-FAR-based contract overview with a specific focus on Other Transaction Authorities (OTA). Using the new Department of Defense (DoD) prototype OTA that was enacted in November 2015, Kuchina-Musina and McMartin conclude by presenting a methodology for examining the effectiveness of OTs. Specifically, they show the way the government is evaluating the DoD statute and discussions on some of the additional guidance the DoD uses to implement this authority. Bringing academic literature on innovation policy and applying it to the practitioner environment, Public Procurement for Innovation provides its audience an understanding of models, methods, and techniques the US uses to promote the development of innovative technologies and products. Clearly written and impeccably researched, the book fills a huge void in the literature on public procurement.

Dolores Kuchina-Musina holds a Ph.D. in Public Administration and Policy and a Master of Business Administration from Old Dominion University. She received her Bachelor of Science in Business Administration from Christopher Newport University. She has been published in the *Journal of Public Procurement, Politics,* and *Policy* and is working on additional research. Dr. Kuchina-Musina has a graduate certificate in Public Procurement and Contract Management (PPCM), is a Certified Federal

Contract Manager (CFCM), and is a National Contract Management Association (NCMA) Fellow. Her research focus includes innovation policy, Other Transaction Authorities, federal procurement reform, and decision-making models. Dr. Kuchina-Musina is the CEO/Founder at REXOTA Solutions, LLC, a consulting company in the DC-Metro area, focusing on strategy, planning, pursuit, capture, management, and administration of federal contract/agreement awards. Dr. Kuchina-Musina has over a decade of experience in federal, state, and international public procurement, specifically directing and managing strategic pursuits in collaboration with business development, proposal development, and change management initiatives for Federal procurement acquisitions. She has also been a strategic partner in providing risk management support for Federal compliance contracts for public and private sector organizations.

Benjamin McMartin, Esq, holds a J.D. from the University of Detroit Mercy School of Law and a Bachelor of Arts from Oakland University in Rochester, Michigan. Mr. McMartin has been published through the George Mason University Center for Government Contracting White Paper Series and has several articles published relating to Federal and international procurement policy. He is a Certified Professional Contracts Manager (CPCM), a National Contracts Management Association (NCMA) Fellow, and a Senior Fellow of the Greg and Camille Baroni Center for Government Contracting at George Mason University. Mr. McMartin's research focus areas include Federal and defense procurement policy, intellectual property, export control, artificial intelligence policy, and emerging technology trends for national security. Mr. McMartin has more than a decade of practical experience as a contracting officer and acquisition manager for the Department of Defense and was instrumental in the development and implementation of defense policy and guidance for the use of Other Transaction Authorities while serving as Chief of the Acquisition Management Office for the US Army Combat Capabilities Development Command – Ground Vehicle Systems Center.

Routledge Research in Public Administration and Public Policy

Behavioural Public Policy in Australia
How an Idea Became Practice
Sarah Ball

Service-Learning for Disaster Resilience
Partnerships for Social Good
Edited by Lucia Velotti, Rebecca M. Brenner and Elizabeth A. Dunn

Complex Governance Networks
Foundational Concepts and Practical Implications
Göktuğ Morçöl

Enhanced Parliamentary Oversight
Promoting Good Governance in Smaller States
Edited by Frederick Stapenhurst, Anthony Staddon, Isabelle Watkinson and Lesley Burns

Public Administration in Hong Kong
Dynamics of Reform and Executive-Led Public Policy
Wei Li

Name, Image, and Likeness Policies
Institutional Impact and States Responses
Darrell Lovell and Daniel J. Mallinson

Public Procurement for Innovation
Research and Development at the US Federal Level
Dolores Kuchina-Musina and Benjamin McMartin

For more information about this series, please visit: www.routledge.com/Routledge-Research-in-Public-Administration-and-Public-Policy/book-series/RRPAPP

Public Procurement for Innovation

Research and Development at the US Federal Level

Dolores Kuchina-Musina and Benjamin McMartin

NEW YORK AND LONDON

First published 2024
by Routledge
605 Third Avenue, New York, NY 10158

and by Routledge
4 Park Square, Milton Park, Abingdon, Oxon, OX14 4RN

Routledge is an imprint of the Taylor & Francis Group, an informa business

© 2024 Dolores Kuchina-Musina and Benjamin McMartin

The right of Dolores Kuchina-Musina and Benjamin McMartin to be identified as authors of this work has been asserted in accordance with sections 77 and 78 of the Copyright, Designs and Patents Act 1988.

All rights reserved. No part of this book may be reprinted or reproduced or utilised in any form or by any electronic, mechanical, or other means, now known or hereafter invented, including photocopying and recording, or in any information storage or retrieval system, without permission in writing from the publishers.

Trademark notice: Product or corporate names may be trademarks or registered trademarks, and are used only for identification and explanation without intent to infringe.

ISBN: 978-1-032-50433-9 (hbk)
ISBN: 978-1-032-50434-6 (pbk)
ISBN: 978-1-003-39845-5 (ebk)

DOI: 10.4324/9781003398455

Typeset in Times
by Apex CoVantage, LLC

We dedicate this book to the international community of innovators, public procurement professionals, and academics, helping pave the way for a better future for all of us.
– Both Authors

I dedicate this book to my husband, Clinton J. Stavrou. Thank you for always believing in my ideas and listening to them, even if it is at two in the morning.
– Dolores Kuchina-Musina

Contents

Foreword	*xiii*
Preface	*xv*
List of Acronyms and Abbreviations	*xvi*

**1 Innovation Policy and Implementation Through
   Public Procurement** — 1
   Why Should We Care About Innovation Policy? 1
   Setting the Groundwork 4
   Summary and Layout of the Book 7

2 Public Procurement and Innovation — 10
   Three Competing Models of Innovation Policy 10
   Overview of Federal Contract Laws and Regulations 15
   *Tensions Between Innovation and Procurement
       Policy 17*

3 Traditional Contract Pathways and Strategies for R&D — 21
   Public Procurement and the Contracting Process 21
   Traditional Methods the Government Contracts Out 25
   Summary 30

4 Alternative Contract Pathways for R&D — 32
   Technology Transfer and Alternative Contract Pathways 32
   Assistance Agreement Instruments 35
   Acquisition Agreement Instruments 37
   Summary 40

**5 Other Transaction Authorities:
   A Department of Defense Case Study** — 43
   Background 43
   Academics Tackling the Subject of Other Transactions 46

Department of Defense Other Transaction Authorities 47
Current Other Transaction Reporting Capabilities 49
Measuring Innovation Policy Effectiveness 52
Summary 61

6 Conclusion and Recommendations 66
Public Procurement and Innovation Policy Considerations 66

Index 71

Foreword

Innovation has been the driving force behind our society's progress for centuries. Nevertheless, pursuing innovation is not a solitary endeavor; it requires the orchestration of multifaceted efforts, the alignment of resources, driven people, and a strategic vision. In this regard, government agencies, particularly the Department of Defense, have often stood at the forefront of innovation, ushering in technological advancements that find their way into civilian applications, enriching the lives of millions. Along this journey, governments have played a pivotal role, acting as both catalyst and curator, guiding the evolution of innovation through policy and procurement. In this remarkable voyage, where innovation meets governance, lies a compelling and educational narrative. It is the narrative that this book, authored by two esteemed experts in innovation and public procurement within the Department of Defense, unveils in vivid detail.

As we delve into the pages of this book, we find ourselves at the nexus of innovation policy and public procurement. This duo has the power to shape industries, elevate nations, and safeguard our shared future.

The road to innovation in the public sector is fraught with unique challenges. It demands a delicate balance between the need for security, accountability, and the imperative to foster innovation. It calls for visionary policy frameworks that nurture creativity while upholding national interests. It requires the efficient management of government contracts to translate ideas into reality. Our authors guide us at this crossroads of ambition and pragmatism, offering insights forged in the crucible of US Federal R&D.

The pages of this book are replete with stories of breakthroughs, struggles, and triumphs. They recount the tireless efforts of scientists, engineers, and policymakers who, armed with a vision of a safer and more technologically advanced future, have driven the innovation engine forward. We learn how public procurement, with its intricate web of regulations and requirements, has been both the lifeblood and the obstacle to progress. Through these narratives, we gain a deeper understanding of the intricate dance between government agencies and the private sector, each reliant on the other to fuel the engine of innovation.

The authors' perspective on innovation policy and public procurement is informed by years of experience. The insights shared in this book provide a roadmap for policymakers, entrepreneurs, and innovators alike, offering a glimpse into the intricacies of navigating the ever-evolving landscape of innovation within the public sector. In a world marked by uncertainty and rapid change, the role of government in fostering innovation has never been more critical. It is a role that extends beyond borders, transcends political ideologies, and unites nations to pursue a better future. It is a role that demands collaboration, vision, and a steadfast commitment to excellence. It is a role that this book explores with depth and clarity, offering valuable insights that will resonate with policymakers, scholars, and practitioners alike.

In closing, I invite you to embark on this intellectual journey through the frontiers of innovation, policy, and government contracts. The following pages are a testament to the power of human ingenuity, the importance of strategic foresight, and the enduring legacy of those who dare to dream beyond the constraints of the known. As we delve into the heart of this narrative, we find ourselves as spectators and active participants in the grand endeavor of shaping a brighter future.

Madelaine Sumera

Preface

Over the past five years, we have worked on many research projects and sought to combine those efforts in this book for anyone to use a desktop guide and point of reference to help expand on this research. In addition, this book was needed when measuring the effectiveness of innovation policy has increased in importance. Institutional knowledge is lost with so many statutes and regulations being drafted annually, but we can capture them and help define the future of public procurement and innovation. This book is a starting point to capture some of the institutional knowledge and serves as a primer for anyone and everyone in this industry.

Thank you to Dr. John C. Morris for encouraging the first step and seeing our vision of this critical topic. Next, we want to acknowledge Sana Hoda Sood and Meghan Ries, who read and reviewed our book during the writing process. We appreciate you cheering us on this year as we embark on this new beginning. We also want to thank everyone on the publishing team. This was our first book together, and all of you made it an enjoyable process. Lastly, we want to thank everyone who made time to talk with us, provide resources, and make intros to experts. The writing process can be difficult and lonely, and we are so grateful to everyone who helped us.

Acronyms and Abbreviations

Key Term	Acronym	Definition
Agreement		The mutually agreed terms and conditions of the parties to an Other Transaction (OT). Absent exceptional circumstances, it will be a legally binding written instrument.
Autoregressive Integrated Moving Average	ARIMA	The Autoregressive Integrated Moving Average model is used as a forecasting tool to predict how something will act in the future based on past performance. It is used in technical analysis to predict an asset's future performance.
Awardee		Any responsible entity that is a signatory to an Other Transaction (OT) agreement. A sub-awardee is any responsible entity performing an effort under the OT agreement other than the awardee.
Bayh-Dole Act		The Bayh-Dole Act or Patent and Trademark Law Amendments Act is a US legislation dealing with inventions from federal government–funded research. Sponsored by two senators, Birch Bayh of Indiana and Bob Dole of Kansas, the Act was adopted in 1980 and is codified at 94 Stat.
Commercial Operation and Support Savings Initiative	COSSI 97	The Commercial Operational and Support Savings Initiative aims to improve readiness and reduce operations and support (O&S) costs by inserting existing commercial items or technology into military legacy systems.
Congressional Research Services	CRS	The Congressional Research Service, or Congress's think tank, is a public policy research institute of the US Congress.
Cooperative Agreement		A cooperative agreement reflects a relationship between the US government and a recipient. It is used when the government aims to assist the intermediary in providing goods or services to the authorized recipient.

Acronyms and Abbreviations xvii

Key Term	Acronym	Definition
Cooperative Research and Development Agreement	CRADA	A Cooperative Research and Development Agreement is any formal written agreement between one or more Federal laboratories and non-federal parties under which the government, through its laboratories, provides personnel, services, facilities, equipment, intellectual property, or other resources.
Cost Accounting Standards	CAS	Cost Accounting Standards are 19 standards and rules promulgated by the US government to determine costs on negotiated procurements. CAS differs from the Federal Acquisition Regulation (FAR) in that FAR applies to substantially all contractors, whereas CAS applies primarily to the larger ones.
Defense Advanced Research Projects Agency	DARPA	The Defense Advanced Research Projects Agency is a research and development agency of the US Department of Defense responsible for developing emerging technologies for use by the military.
Defense Contract Audit Agency	DCAA	The Defense Contract Audit Agency is an agency of the US Department of Defense under the direction of the Under Secretary of Defense. It was established in 1965 to perform all contract audits for the Department of Defense.
Defense Contract Management Agency	DCMA	The Defense Contract Management Agency is an agency of the US federal government reporting to the Under Secretary of Defense for Acquisition and Sustainment. It is responsible for administering contracts for the Department of Defense and other authorized federal agencies.
Defense Contract Management Agency	DCMA	The Defense Contract Management Agency is an agency of the US federal government reporting to the Under Secretary of Defense for Acquisition and Sustainment. It is responsible for administering contracts for the Department of Defense and other authorized federal agencies.
Defense Federal Acquisition Regulation Supplement	DFARS	The Defense Federal Acquisition Regulation Supplement to the Federal Acquisition Regulation (FAR) is administered by the Department of Defense (DoD). The DFARS implements and supplements the FAR. The DFARS contains requirements of law, DoD-wide policies, delegations of FAR authorities, deviations from FAR requirements, and policies/procedures that significantly affect the public.

Acronyms and Abbreviations

Key Term	Acronym	Definition
Department of Defense	DoD	The US Department of Defense is an executive branch department of the federal government charged with coordinating and supervising all agencies and functions of the government related to national security and the US Armed Forces.
Department of Energy	DOE	The US Department of Energy is a cabinet-level department of the US government concerned with the US policies regarding energy and safety in handling nuclear material.
Department of Health and Human Services	HHS	The US Department of Health and Human Services, also known as the Health Department, is a cabinet-level executive branch department of the US federal government with the goal of protecting the health of all Americans and providing essential human services.
Department of Homeland Security	DHS	The US Department of Homeland Security is the US federal executive department responsible for public security, roughly comparable to other countries' interior or home ministries.
Department of Transportation	DOT	The US Department of Transportation is a federal cabinet department of the US government concerned with transportation. It was established by an act of Congress on 15 October 1966, and began operation on 1 April 1967. The US Secretary of Transportation governs it.
Educational Partnership Agreements	EPA	Educational Partnership Agreements aim to foster and elevate the study of scientific disciplines at all educational levels. These agreements constitute a formal pact between a defense laboratory and an educational institution to facilitate the transfer and improvement of technology applications and deliver technological assistance across the education spectrum from pre-kindergarten onward.
Federal Acquisition Regulations	FAR	The Federal Acquisition Regulations cover many of the US military and NASA contracts. The largest single part of the FAR is Part 52, which contains standard solicitation provisions and contract clauses.
Federal Acquisition Streamlining Act of 1994 (PL 103-355)	FASA	On 13 October 1994, President Clinton signed the Federal Acquisition Streamlining Act of 1994 into law. The comprehensive acquisition reform legislation streamlines the federal government's acquisition system and dramatically changes the way the government performs its contracting functions.

Acronyms and Abbreviations xix

Key Term	Acronym	Definition
Federal Grant and Cooperative Agreement Act of 1977 (PL 95-224)		The Federal Grant and Cooperative Agreement Act of 1977 (PL 95-224) guided government agencies in deploying Federal funds – particularly by distinguishing between contracts, cooperative agreements, and grants. This act addressed US Congressional concerns over the perceived misuse of assistance agreements to circumvent competition and procurement rules.
Federal Procurement Data System – Next Generation	FPDS-NG	The Federal Procurement Data System is a single source for US government-wide procurement data. The Federal Procurement Data Center, part of the US General Services Administration, manages the Federal Procurement Data System, which is operated and maintained by IBM.
Federal Technology Transfer Act (PL 99-502)		The Federal Technology Transfer Act (PL 99-502), codified in 15 USC 3710, was passed by Congress in 1986 as an amendment to the Stevenson-Wydler Act of 1980. Its primary objective is to enhance industry access to technologies from federal laboratories. This legislation led to establishing a nationwide network of federal laboratories, agencies, and research centers called the Federal Laboratory Consortium for Technology Transfer.
Fiscal Year	FY	A fiscal year is used in government accounting, which varies between countries, and for budget purposes. It is also used for financial reporting by businesses and other organizations. The time spans from 1 October to 30 September of every year.
Information Asymmetry		In contract theory and economics, information asymmetry deals with the study of decisions in transactions where one party has more or better information than the other. Information asymmetry creates an imbalance of power in transactions, which can sometimes cause the transactions to be inefficient, causing market failure in the worst case.
Innovation Policy		For this book, innovation policy is any policy that promotes innovation.
Government Accountability Office	GAO	The US Government Accountability Office is a legislative branch government agency that provides auditing, evaluation, and investigative services for the US Congress. It is the supreme audit institution of the US federal government.

Acronyms and Abbreviations

Key Term	Acronym	Definition
Grant		A grant is how the government funds your ideas and projects to provide public services and stimulate the economy. Grants support critical recovery initiatives, innovative research, and many other programs listed in the Catalog of Federal Domestic Assistance (CFDA).
National Aeronautics and Space Administration	NASA	The National Aeronautics and Space Administration is an independent agency of the US federal government responsible for the civilian space program and aeronautics and space research. NASA was established in 1958, succeeding the National Advisory Committee for Aeronautics.
National Defense Authorization Act	NDAA	The National Defense Authorization Act is the name for each of a series of US federal laws specifying the annual budget and expenditures of the US Department of Defense. The first NDAA was passed in 1961.
National Technology Transfer and Advancement Act of 1995 (PL 104-113)	NTTAA	The National Technology Transfer and Advancement Act was signed into law on 7 March 1996. By enacting the NTTAA into law, Congress found it could promote economic, environmental, and social well-being by bringing technology and industrial innovation to the marketplace.
Nontraditional Defense Contractor	NDC	An entity that is not currently performing and has not performed, for at least the one year preceding the solicitation of sources by Department of Defense for the procurement or transaction, any contract or subcontract for the DoD that is subject to full coverage under the cost accounting standards prescribed pursuant to section 1502 of title 41 and the regulations implementing such section (see 10 USC 3014).
North American Industry Classification System	NAICS	The North American Industry Classification System is the standard used by Federal statistical agencies in classifying business establishments to collect, analyze, and publish statistical data related to the US business economy.
Office of the Secretary of Defense	OSD	The Office of the Secretary of Defense is a headquarters-level staff of the US Department of Defense.
Other Transaction	OT	Refers to any transaction other than a procurement contract, grant, or cooperative agreement.

Key Term	Acronym	Definition
Other Transaction Authority	OTA	Refers to the statutory authorities that permit a US federal agency to enter into transactions other than contracts, grants, or cooperative agreements.
Partnership Intermediary Agreements	PIA	A Partnership Intermediary Agreement is a contract, agreement, or memorandum of understanding with a non-profit partnership intermediary to engage academia and industry on behalf of the government to accelerate tech transfer and licensing.
Public Procurement		Public procurement refers to the process by which public authorities, such as government departments or local authorities, purchase work, goods, or services from companies.
Principal-Agent Theory		Principal agent theory describes the challenges when an entity, the "agent," represents another entity, known as the "principal."
Procurement Contract		A contract awarded according to the Federal Acquisition Regulation.
Product Service Codes	PSC	Also referred to as the US government uses federal supply codes, product service codes to describe the products, services, and research and development purchased by the government. Government procurement specialists and government contractors alike require a solid understanding of these codes to produce quality partnerships between buyers and suppliers.
Public Private Partnerships	PPP	A public-private partnership is a cooperative arrangement between two or more public and private sectors, typically of a long-term nature. In other words, it involves government and business that work together to complete a project and/or to provide services to the population.
Request for Proposal	RFP	Request for Proposals are one type of solicitation government agencies use.
Research and Development	R&D	Research and development occurs when private companies gather knowledge to create new products or discover new ways to improve their existing products and services. Larger companies may have their own research and development team that will test and refine products or processes before commercial use. The Department of Defense defines this by the Research, Development, Test, and Evaluation (RDT&E) budget activity.

Key Term	Acronym	Definition
Science and Technology	S&T	Science and technology is an interdisciplinary topic encompassing science, technology, and their interactions: science is a systematic enterprise that builds and organizes knowledge through explanations and predictions about nature and the universe.
Small Business Innovation Research Program	SBIR	This Small Business Innovation Research program aims to stimulate technological innovation, prioritize small businesses to meet US research and development needs, and increase private sector commercialization of innovation funded by the US government. The SBIR program is codified at § 9 of the Act, 15 USC 638.
Small Business Technology Transfer Program	STTR	The Small Business Technology Transfer program's statutory purpose is to stimulate a partnership between innovative small businesses and non-profit research institutions, which supports the goals for commercializing innovative technologies. The STTR program is also codified at § 9 of the Act, 15 USC 638.
Solicitation		When the government wants to buy a good or service, it issues a solicitation. Solicitations are documents that clarify the government's requirements so that businesses can submit competitive bids.
Stevenson-Wydler Technology Innovation Act of 1980 (PL 96-480)		The Stevenson-Wydler Technology Innovation Act was signed into law by US President Jimmy Carter on 21 October 1980. The primary focus of the Stevenson-Wydler Act was to disseminate information from the federal government to the public and require federal laboratories to actively engage in the technology transfer process.
Technology Investment Agreement	TIA	Technology Investment Agreements are assistance instruments used to stimulate or support research. The goal for using TIAs, like other assistance instruments used in defense research programs, is to foster the best technologies for future defense needs. TIAs differ from and complement other assistance instruments available to agreements officers, in that TIAs address the goal by fostering civil-military integration.

Key Term	Acronym	Definition
Transaction		The entire process of interactions related to, entering into an agreement, executing, and transitioning a prototype project.
United States Code	USC	A comprehensive body of laws passed by Congress and organized topically under 50 titles. A typical citation to the code (e.g., 15 USC § 290) gives the title number (a number from 1 to 50), the abbreviated title of the code itself (USC), and the section number under which the statute may be found.

1 Innovation Policy and Implementation Through Public Procurement

Why Should We Care About Innovation Policy?

Innovation policy plays a crucial role in shaping the progress of societies, economies, and organizations. It refers to a set of strategies and initiatives designed to foster and support innovation, enabling the creation and diffusion of new ideas, technologies, and processes. By supporting research and development, promoting entrepreneurship, and incentivizing cross-sector collaboration between academia, industry, and government, innovation policy stimulates the creation of new industries, jobs, and markets. Innovation policy enables governments to adapt to changing circumstances, maintain a competitive edge, and attract investment and talent. In addition, it addresses a wide array of societal challenges from healthcare to energy, from transportation to education. Innovation has the potential to tackle complex problems and to enhance well-being. Through targeted policies, governments can encourage innovation in areas such as sustainable technologies, healthcare advancements, and social modernization, leading to improved services, increased efficiency, and better access to resources for citizens.

By supporting research and development, providing funding mechanisms, and promoting efficient protection of intellectual property rights, policy frameworks encourage individuals and organizations to explore new ideas, take risks, and pursue breakthrough technological developments. Fostering an environment of creativity and risk acceptance encourages entrepreneurship, enhances productivity, and promotes continuous learning and adaptation, all crucial attributes in an ever-changing world. Moreover, innovation policy is pivotal in addressing societal inequalities and promoting inclusivity. By ensuring equitable access to resources, education, and opportunities, innovation policy can bridge the digital divide, reduce resource disparities, and promote social mobility. Policies can encourage diversity and inclusion in innovative ecosystems, leading to a wider range of perspectives, ideas, and solutions. As such, understanding and prioritizing innovation policy are essential for governments tasked to keep pace with an ever-changing workforce, fluctuations in available funding, and the rapid increase in the pace of technological advancement.

DOI: 10.4324/9781003398455-1

Changing Workforce

Over the past century, the US workforce has experienced transformative shifts driven by globalization, technological advancements, economic fluctuations, and societal trends. These transformations have reshaped the nature of work, demanded new skills, and altered the employer-employee relationship.

Over the past 20 years, globalization has significantly increased, leading to outsourcing labor-intensive tasks to countries with lower costs. Manufacturing jobs have shifted from developed to developing nations, impacting employment opportunities in some regions. At the same time, globalization has opened new markets, enabling businesses to expand internationally, and creating new avenues for job growth and innovation. In combination with globalization, the need for new technological developments has been on the rise. In the last two decades alone, we have seen several technological revolutions, including automation, artificial intelligence, and digitalization – all of which have revolutionized the workforce. These advances have led to the displacement of manual and repetitive tasks while creating new job opportunities in technology-driven sectors. Due to these two forces, it is no surprise that education and professional skill requirements have evolved significantly. Traditional industrial skills have given way to a demand for diverse skills, including technological proficiency, critical thinking, and problem-solving. Continuous upskilling and reskilling are now necessary for individuals to remain employable and adaptable, especially when organizations must also actively support diversity and inclusion efforts, promote equal opportunities, combat biases, and remove barriers to entry and advancement.

Remote and flexible work arrangements have gained prominence, accelerated by the COVID-19 pandemic. Organizations and employees have recognized the viability and benefits of remote work, necessitating a reimagining of the traditional office-based work environment. These changing work arrangements and economic pressures surrounding them have also resulted in the rise of the gig economy, which has had a significant shift in employment structures, with an increase in freelancers, independent contractors, and temporary workers.

Decreased US Federal Investment in Innovation

In 1960, the US accounted for 69% of global research and development (R&D), with US defense-related R&D alone accounting for more than one-third of global R&D. During this time, the US federal government funded approximately twice as much R&D as the private sector. According to John F. Sargent Jr., Marcy E. Gallo, and Moshe Schwartz, "[F]rom 1960 to 2016, the US share of global R&D fell to 28%, and the US federal government's share of total US R&D fell from 65% to 24%, while business' share more than doubled from 33% to 67%" (Sargent Jr., Gallo, & Schwartz, 2018, p. 7).

Given that the US government has historically driven national innovation by serving as the largest global investor and the primary US investor in higher-risk emerging technologies, these declining trends raise concerns. The success of many of these higher-risk government ventures is hard to overstate. Government investments "have proved transformative, creating entirely new markets and sectors, including the Internet, nanotechnology, biotechnology, and clean energy" (Mazzucato, 2015, p. 4). These trends are causing US policymakers to become increasingly concerned about the relationship between the declining government investments in innovation and its impact on US competitiveness in global markets and in national security.

Lawmakers are further concerned about the relationship between Federal procurement laws, regulations, and policies that impact US investment in emerging technology sectors. Such laws, regulations, and policies have been shown to have an adverse effect on the ability of agencies to fund technology investment and negatively impact the ability of new technology firms to enter the federal marketplace. The federal acquisition system has created cumbersome administrative barriers for technology firms seeking to enter the market, disincentivizing engagement with the federal and defense markets. This is a concern, as 67% of the investments in R&D to develop new technologies reside in the private sector. Simply put, the companies spending on the next generation of technology development need help navigating the federal procurement process, and thus, their R&D investments concentrate on purely commercial applications.

Technology Increasing at a Rapid Rate

The last two decades have witnessed an unprecedented and exponential increase in technological advancements, transforming nearly every aspect of our lives. Technology has experienced significant growth and impact, particularly in three key areas: communication and connectivity, artificial intelligence (AI) and automation, and the proliferation of mobile devices.

Regarding communication and connectivity, the advent of the Internet in 1983 and the subsequent growth of broadband and wireless technologies have revolutionized how we interact and share information. The widespread adoption of smartphones and the availability of high-speed internet have connected people across the globe, enabling instant communication, access to vast amounts of information, and the rise of social media platforms. The Internet has become essential for business, education, entertainment, and social interactions, creating new opportunities and transforming industries.

Artificial intelligence and automation have made remarkable progress over the past 20 years, empowering machines to perform tasks that previously required human intelligence. Machine learning algorithms and deep learning techniques have fueled breakthroughs in AI, enabling applications such as natural language processing, image recognition, and autonomous systems. Automation has profoundly impacted various industries, from manufacturing and logistics to customer service and healthcare. Robotic process automation (RPA) and the integration of AI-driven systems have improved efficiency, accuracy, and productivity while raising questions about the future of work and the potential for job displacement.

The proliferation of mobile devices, particularly smartphones, has transformed how we access information, communicate, and conduct business. The miniaturization of technology, coupled with increased computing power, has put powerful devices in the hands of billions of people worldwide. Mobile apps have become integral to our daily routines, offering services for communication, entertainment, shopping, transportation, and more. The ubiquity of smartphones has created new business models and opportunities for innovation while presenting challenges related to privacy, cybersecurity, and digital well-being.

Setting the Groundwork

When people think of innovation, they associate the term with research and development. This connection is evident from examining academic and practitioner literature and how the US government tracks its budgetary dollars. The *Oxford Dictionary* defines research and development as "work directed toward the innovation, introduction, and improvement of products and processes." In addition to research and development, the terms *innovation* and *technology* must also be defined for this book. Innovation is a new method, idea, or product; in other words, innovation can be either a product or a process (Borrás & Edquist, 2013). Technology is the application of new knowledge and innovation from research and development efforts. The link between these three terms is vital because R&D leads to innovation and technological change (Link, 2006; Vonortas, Rouge, & Aridi, 2014). The necessity to manage innovation and technological change has thus become an essential consideration for policymakers. The declining trend in funding R&D is troubling because R&D efforts are proven to lead to innovations like the internet and cell phones (Link, 2006). Public procurement and innovation are inextricably linked because "public procurement can be used as a tool to stimulate innovation resulting in a new type of product or to open the marketplace for innovation without a specific product" (Kuchina-Musina & Morris, 2022, p. 7).

Defining Innovation Policy

Innovation policy combines systems and organizations that may directly or indirectly affect innovation (Kuchina-Musina & Morris, 2022). *Innovation policy* is a relatively new term when discussing the policymaking agenda and has gained some traction over the last decade (Edler & Fagerberg, 2017).

The term *innovation policy* has historically gone under many different labels, "such as industrial policy, science policy, research policy, or technology policy" (Edler & Fagerberg, 2017, p. 5). For the purposes of this book, innovation policy is any policy that promotes innovation. This definition includes policies that directly support innovation using funding mechanisms such as grants, contracts, or indirect support such as incentive tax programs for the private sector matching the private firm's expenditure with public funding (Vonortas et al., 2014). Understanding innovation policy has become increasingly important. For example, "from the 1970s onward, Douglas North, Robert Thomas, Nathan Rosenberg, and other economic historians argued that innovation was aided by specific government institutions and policies" (Taylor, 2016, p. 307). The decrease in federal funding for R&D can be an indicator of the actions of public agencies. These combined actions of public agencies, whether direct or indirect, are innovation policy because they affect innovation in one way or another (Borrás & Edquist, 2013). Understanding the impact of innovation policy is crucial to help stimulate policies that address societal challenges such as cybersecurity, climate change, unemployment, and inequality.

Defining Public Procurement

The topic of *public procurement*, especially contracting, has been gaining considerable attention because of the pressures of budget cuts and the demand for increased efficiencies in the public sector (Jurisch, Ikas, Wolf, & Krcmar, 2013). The private sector seeks efficiency and depends on competition to thrive, while the government is driven by budget requirements and resource allocation (Kettl, 1993; Jurisch et al., 2013). Public procurement is essential to support the demand for valuable and cost-saving services. The goals of public procurement include reducing cost, increasing quality, timeliness, risk management, accomplishing social and economic objectives, and maximizing competition while maintaining integrity and transparency (Pitzer & Thai, 2009). To achieve these goals, procurement departments must ensure proper leadership, management, and capability of the procurement service. Achieving these goals, however, requires time, effort, and cost; thus, there is a trade-off between transparency, timeliness, and efficiency in public procurement (Pitzer & Thai, 2009). In addition to these practical implications, procurement departments face external social, economic, and political goals and challenges. For example, during every election and change in administration, a procurement department may have different initiatives, metrics, and policies to abide by making it difficult to ensure consistency year after year.

These problems are magnified in a public procurement relationship because of the complex nature of goods and services to be procured by the government. Moreover, public procurement occurs in complex organizational settings that have numerous policies and procedures along with ever-changing requirements driving the prioritization for the agency. The nature of the procurement relationship creates both the likelihood of a principal-agent problem and information asymmetry that can affect large procurement programs, organizational culture, future policy, and public-private relationships. Thus, how an organization accomplishes its procurement goals and the decision-making processes that are used within the organizational settings inherent in its process can be disparate and fragmented, absent clear national policies and regulations.

Innovation Policy in the US

Innovation policy allows the market to provide a solution in the form of a good or service, stimulate the environment, and promote learning. With a demand-driven market, public procurement can trigger innovation by satisfying a human need and solving a societal problem (Edquist & Zabala-Iturriagagoitia, 2012). This approach can also mitigate some trade-offs, such as risk, and improve quality for complex procurement requirements and is directly linked to how procurement requirements are analyzed.

Innovation policy should be flexible and simple. It should allow for innovative solutions that are not limited by strict limitations that may hinder competition and the incentive for innovation. This creates a burden on organizations and discourages the private sector from participating in R&D efforts with the government. One suggested solution is to deviate from the traditional public procurement process and develop a process for R&D efforts. This solution, however, again creates an arduous task that is overly cumbersome and expensive. Procurement officials can continue to promote collaboration to adopt the proper incentives and risk-sharing in the contractual relationship (Carbonara & Pellegrino, 2018). In addition, procurement officials must be ready to take feedback and apply it to actionable change to continue promoting innovative solutions instead of becoming a barrier. Finding the proper structure for the organization depends on the organization's culture and mission.

Currently, the US has established neither a single national policy on innovation nor a national innovation strategy. Instead, a patchwork of strategies related to critical and emerging technologies has emerged, which seek innovation within defined technology areas. For example, in April 2023, the White House Office of Science and Technology Policy, in coordination with the US Department of Energy and the US Department of State, released a National Innovation Pathway of the United States, which documented a US clean energy innovation strategy (National Innovation Pathway, 2023). Further, in May 2023, the White House released the US Government National Standards Strategy for Critical and Emerging Technology. This strategy aims to "accelerate standards efforts led by the private sector to facilitate global markets, contribute to interoperability, and promote U.S. competitiveness and innovation" (The United States Government, 2023). The standards strategy aligns with the US National Security Strategy, which states that the "United States is pursuing a modern industrial and innovation Strategy" (The White House, 2022a, p. 14). In addition, large investments have been made in recent years in increasing innovation in critical and emerging technologies. The CHIPS + Science Act, signed into law in August 2022, established a technology, innovation, and partnerships directorate within the National Science Foundation; authorized $10 billion to invest in regional innovation; authorized investment in science, technology, engineering, and math (STEM) education and innovation; and authorized $1.5 billion to promote US innovation in wireless supply chains (The White House, 2022b). In contrast, the European Union (EU) has established an innovation policy as a subset of its economy and industrial, energy, and research policies. As defined by the EU, innovation policy is the "interface between research and technological development policy and industrial policy" (European Parliament, 2019). It seeks to create a framework conducive to bringing ideas to market. Although the US has not settled on a single innovation policy, the connection between public procurement and innovation policies is consistent. The US National Security Strategy states that the US is "using public procurement in critical markets to stimulate demand for innovation" (The White House, 2022a, p. 14).

Summary and Layout of the Book

This chapter is a high-level introduction to the research and the innovation policy environment in the US. After reading this chapter, you should be more familiar with current trends in the workforce, the decrease trends of public investment in R&D, and how technology is increasing rapidly by private investment. In addition, you should now have a baseline on key terms and subjects such as innovation policy and public procurement.

The book should serve as a primer or desktop guide to help navigate the robust and extensive US public procurement concepts, as well as how these concepts support innovation. This subject matter is becoming more relevant as global initiatives increase to address rapid change and challenges while supporting technological advancements at scale. This book consists of five analytical chapters and a concluding chapter that discusses the implications of the findings for practitioners and researchers.

Chapter 2 explores how public procurement and innovation policy are tied together. The analysis of this chapter will review current literature in the topic area and some policy mechanisms used in the US. Chapters 3 and 4 will detail Federal Acquisitions Regulation (FAR) and non-FAR procurement pathways and strategies to support innovation. These chapters will provide details about each procurement strategy, enabling the reader to understand the alignment of varying procurement strategies with desired innovation outcomes. The chapter concludes with an introduction of Other Transaction Authorities (OTAs), which are a current focus within federal procurement due to the perceived benefits in flexibility these contracts provide to government and industry. Chapter 5 will present a methodology for examining the effectiveness of Other Transactions using the new Department of Defense prototyping OTA enacted in November 2015. Finally, Chapter 6 concludes the book with a discussion of the implications of the research findings for policymakers, practitioners, and academics interested in the intersection of public procurement and innovation policy.

We hope you enjoy the read.

References

Borrás, S., & Edquist, C. (2013). The choice of innovation policy instruments. *Technological Forecasting and Social Change, 80*(8), 1513–1522.

Carbonara, N., & Pellegrino, R. (2018). Fostering innovation in public procurement through public private partnerships. *Journal of Public Procurement, 18*(3), 257–280. doi:10.1108/JOPP-09-2018-016

Edler, J., & Fagerberg, J. (2017). Innovation policy: What, why, and how. *Oxford Review of Economic Policy, 33*(1), 2–23. doi:10.1093/oxrep/grx001

Edquist, C., & Zabala-Iturriagagoitia, J. M. (2012). Public procurement for innovation as mission-oriented innovation policy. *Research Policy, 41*(10), 1757–1769.

European Parliament. (2019, April). Innovation policy | Fact Sheets on the European Union. *Europa.eu*. Retrieved from www.europarl.europa.eu/factsheets/en/sheet/67/innovation-policy

Jurisch, M., Ikas, C., Wolf, P., & Krcmar, H. (2013). Key differences of private and public sector business process change. *e-Service Journal, 9*, 3–27. doi:10.2979/eservicej.9.1.3

Kettl, D. F. (1993). *Sharing power public governance and private markets*. Washington, DC: The Brookings Institution.

Kuchina-Musina, D., & Morris, J. C. (2022). Buying innovation: An examination of public-private partnerships and the decision process for contracting out innovation. *Politics & Policy, 50*(3), 503–515. doi:10.1111/polp.12464

Link, A. N. (2006). *Public/private partnerships: Innovation strategies and policy alternatives*. Springer Science & Business Media.

Mazzucato, M. (2015). *The entrepreneurial state: Debunking public vs. private sector myths* (Vol. 1). London, UK: Anthem Press.

National Innovation Pathway of the United States. (2023). Retrieved from www.whitehouse.gov/wp-content/uploads/2023/04/US-National-Innovation-Pathway.pdf

Pitzer, J. P., & Thai, K. V. (2009). *Introduction to public procurement*. Herndon, VA: National Institute of Governmental Purchasing.

Sargent Jr., J. F., Gallo, M. E., & Schwartz, M. (2018). *The global research and development landscape and implications for the Department of Defense (R45403)*. Washington, DC: Congressional Research Service.

Taylor, M. Z. (2016). *The politics of innovation: Why some countries are better than others at science and technology*. Oxford University Press.

The United States Government. (2023, May 8). Fact sheet: Biden-Harris Administration Announces national standards strategy for critical and Emerging Technology.

The White House. Retrieved from www.whitehouse.gov/briefing-room/statements-releases/2023/05/04/fact-sheet-biden-harris-administration-announces-national-standards-strategy-for-critical-and-emerging-technology/

Vonortas, N. S., Rouge, P. C., & Aridi, A. (2014). *Innovation policy: A practical introduction*. Springer.

The White House. (2022a). National security strategy. *Whitehouse.gov*. Retrieved from www.whitehouse.gov/wp-content/uploads/2022/10/Biden-Harris-Administrations-National-Security-Strategy-10.2022.pdf

The White House. (2022b, August 9). FACT SHEET: CHIPS and science act will lower costs, create jobs, strengthen supply chains, and counter China. *The White House*. Retrieved from www.whitehouse.gov/briefing-room/statements-releases/2022/08/09/fact-sheet-chips-and-science-act-will-lower-costs-create-jobs-strengthen-supply-chains-and-counter-china/

2 Public Procurement and Innovation

Three Competing Models of Innovation Policy

In the 2016 Annual Industrial Capabilities Report to Congress, the Department of Defense acknowledged that it should take advantage of the private sector's rapid growth by leveraging innovation created by "nontraditional defense contractors." In a similar study, a Government Accountability Office (GAO) report, GAO-17-644, in July 2017 used data collected from 12 innovative companies that do not engage in business with the Department of Defense. This report identified six challenges that deter these companies from doing business with the agency. These challenges included 1) complexity of the Department of Defense's process, 2) the unstable budget environment, 3) long contracting timelines, 4) intellectual property rights concerns, 5) government-specific contract terms and conditions, and 6) the inexperienced Department of Defense contracting workforce. The Department of Defense recommended promoting initiatives to make the acquisition process more accessible and flexible (Sullivan, 2017). Such initiatives included several innovation policies, such as the FY 2016 and 2017 NDAA provisions for the Department of Defense, including the codification of the Department of Defense's Other Transaction Authority (10 USC 4022, formerly 10 USC 2371b) for prototypes. This innovation policy authorized the Department of Defense to award follow-on production contracts for successful prototypes without using competitive procedures. It is important to note that although it is beneficial to have all these initiatives and innovation policies to support a solution for the challenges listed earlier, the real problem becomes analyzing the outcomes of such initiatives to find if these policies promote innovative technologies and products.

Before analyzing the outcomes of innovation policies, it is vital first to understand the US government's level of involvement in promoting innovative technologies and products. Three policy paradigms can be used to examine the US approaches to innovation policy: the market, mission, and cooperative models (Bozeman, Crow, & Tucker, 1999). Although these three models were developed for studying research and development policy, they are still applicable when discussing innovation policy due to the natural crossover (Borrás & Edquist, 2013;

Edler & Fagerberg, 2017). The models depicted are based on the level of involvement the government should have in regulating and influencing private sector behavior (Bozeman et al., 1999). The role the government plays determines the model applicable to the specific paradigm. The market model applies "market failure" as the reason the government sponsors research and development (R&D) efforts. The mission model assumes that the government should perform the innovative services of specific missions that cannot be efficiently served by the private sector (e.g., defense and national security-related innovation). Lastly, the cooperative model is a relatively newer model in which the government has a more active role in performing research and developing innovative technology and products for private sector consumption or by merely being a funding vehicle for R&D efforts.

The challenges presented in the General Accountability Office's report GAO-17-644 demonstrate the numerous problems the Department of Defense has encountered, which prevent the Department of Defense from attracting innovative contractors to collaborate and develop new national defense technologies. Applying these three models can help explain the importance of the suggested initiatives in the aforementioned report and how innovation policies influence the Department of Defense's contracting activities.

Before analyzing the outcomes of innovation policies, it is vital first to understand the US government's level of involvement in promoting innovative technologies and products. Three policy paradigms can be used to examine the US approaches to innovation policy: the market, mission, and cooperative models. Although these three models were developed for studying R&D policy, they are still applicable when discussing innovation policy because innovation policy includes R&D policy (Borrás & Edquist, 2013; Edler & Fagerberg, 2017). These three models (see Table 2.1) are based on the level of involvement the government should have in regulating and influencing private sector behavior (Bozeman et al., 1999). The market model applies market failure as the reason the government sponsors R&D efforts. The mission model assumes that the government should perform the innovative services of specific missions that cannot be efficiently served by the private sector (e.g., defense and national security-related innovation). Lastly, the cooperative model is a new model in which the government has a more active role in performing research and developing innovative technology and products for private sector consumption or by merely being a funding vehicle for R&D efforts.

The 1980s in the US were a time of economic uncertainty, leading to scholars re-examining the private sector's role in innovation. The market failure paradigm began to lose its luster, and scholars needed a new way of assessing innovation policy. The cooperative model was the challenger to the market failure paradigm, which emphasized cooperation among sectors (Bozeman et al., 1999). This is important to note in the innovation policy timeline as the Department of Defense assessed its challenges in developing innovative

Table 2.1 The Competing Innovation Policy Models

	Market Model	Mission Model	Cooperative Model
Core Assumptions	1. Reliance on markets as the most efficient allocator of information and technology. 2. The government's role is constrained to factors such as extensive externalities, high transaction costs, and information distortions. Small, mission domain, chiefly in defense. 3. Minimal government role in innovation. 4. Innovation flows from and to the private sector.	1. The government's role should be closely tied to the authorized programmatic missions of agencies. 2. Government R&D is limited to missions of agencies but not confined to defense. 3. The government should not compete with the private sector in innovation and technology but should have a connection with the traditional activities of the agencies.	1. Markets are not always the most efficient route to innovation and economic growth. 2. The global economy requires more centralized planning and broader support for civilian technology development. 3. Government can play a role in developing technology, especially pre-competitive technology, for use in the private sector.
Peak Influence	Applicable to all periods	1945–1965; 1992–present.	1992–1994; 2010–present.
Policy Examples	De-regulation; contraction of government role; R&D tax credits; capital gains tax roll back. Little or no need for federal laboratories except in defense support.	Creation of energy policy R&D, agricultural labs, and other broad mission frameworks.	Expansion of federal laboratory roles in technology transfer and cooperative research; manufacturing extension policies.
Theoretical Roots	Neoclassical economics.	Traditional liberal governance with a broad definition of government role.	Industrial policy theory.

technologies and products. As aforementioned, Congress enacted Section 251 of Public Law No: 101-189, codified in 10 USC 2371 in 1989, now codified in 10 USC 4021, which provided the Defense Advanced Research Projects Agency (DARPA) the authority to conduct research and technology developments using cooperative agreements and Other Transactions. At the time, this authority was only granted for a two-year pilot program. In 1991, the National Defense

Authorization Act of FY1992 extended this authority to other Department of Defense agencies and made it permanent. This was the first major expansion of authority for other transactions because it enabled other departments in the Department of Defense to use other transactions for their efforts.

The Department of Defense's procurement contract method of promoting innovation is an excellent example of the market failure model and the mission model. The market failure paradigm has been the dominant model for innovation policy by assuming that if there is a need for innovation, then the private sector will "sense the need and respond in an economically efficient manner" (Bozeman et al., 1999, p. 6). Thus, for any non-mission-specific contracts, the Department of Defense applies the market failure paradigm. The mission model becomes applicable for contracts that are mission-specific contracts (e.g., weapons systems). The mission paradigm assumes that the agency mission should drive innovation policy resulting in innovative products. These two models apply to the procurement contract method because they are not significantly different and are the most used innovation policy models.

The procurement contract method is a top-down regulated approach to procuring goods and services from the private sector. A procurement contract is a contract awarded according to the Federal Acquisition Regulation (FAR). Regulations governing the procurement contract method (i.e., the FAR, Cost Accounting Standards (CAS), and the Bayh-Dole Act) are adopted by any government agency that desires to do business with the private sector. This procurement contract approach should be familiar to scholars accustomed to the Mazmanian and Sabatier paradigm of the "single-authority top-down" methodology (Hjern & Hull, 1982). In this approach, "implementation is the carrying out of a basic policy decision, usually made in a statute" (Mazmanian & Sabatier, 1981, p. 540). Government support is proposed once a market failure is identified (Mazzucato, 2015). Later this government support is evaluated as an investment, and value is assessed using cost/benefit analysis. The six challenges identified in GAO-17–644 apply to the procurement contract method, another example of the top-down model.

From 1989 onward, as the cooperative model's popularity grew, the innovative policy promoting other transactions as an alternative contract method evolved. Two years later, the National Defense Authorization Act (NDAA) of FY1994 Public Law No: 103-160 expanded the other transaction authority for Defense Advanced Research Projects Agency (DARPA) to include prototype projects related to weapons or weapons systems procured or developed for a Department of Defense agency. This was significant because the extension of authority allowed other transactions as a procurement method rather than a stimulant for research efforts. This legislation is the second major expansion of authority for other transactions because it changed how the Department of Defense used other transactions. Three years later, the National Defense Authorization Act of FY1997 Public Law No: 104-201 extended Section 845 prototype authority to the remainder of the Department of Defense.

Over the next two decades, the authority of other transactions was both expanded and restrained by enacting additional clarifications and reporting requirements for transparency and accountability of Department of Defense agencies. These reporting requirements included annual reports that would include a description of the transaction, the reason for not using a contract or grant to support the research, the amount of the payment, and other requirements.

The next major expansion of authority for other transactions occurred nearly 20 years after the first major event, with the National Defense Authorization Act of FY1994 (PL 103-160). In 2014, the National Defense Authorization Act for FY2015 (PL 113-291) expanded other transaction authority to include prototypes that were:

> directly related to enhancing the mission effectiveness of military personnel and supporting platforms, systems, components, or materials proposed to be acquired or developed by the Department of Defense, or improvements of platforms, systems, components, or materials in use by the Armed Forces (p. 138).

Before this, other transactions could only be used for weapons and weapons systems.

In 2015, the National Defense Authorization Act of FY2016 (PL 114-92) codified prototype other transactions in 10 USC 4022, thereby rescinding the authority under Section 845, redefining and codifying nontraditional defense contractors in 10 USC 2302(9) now codified in 10 USC 3014 and expanding follow-on production (10 USC 4022(f) formerly in 10 USC 2371b(f)). This legislation is the fourth major expansion of authority, and the focus of this study is on how this innovation policy has promoted the development of innovative technologies and products. To highlight the importance of this study, it is important to review key previous research efforts over the past few decades, which are discussed in the following section.

The amount of investment depends on available funding, and with the trends of the past several decades, the decline in funding for innovation has caused great concern. This is where the mission model and market model are challenged. Available financing combined with the Department of Defense's problem of not attracting innovative firms can create a call for action. These challenges and the need for public and private sectors to work together illustrate the cooperative model's assumption that the market is not always the most efficient way of promoting innovation.

For this reason, the cooperative model has gained interest, especially in the application in the commercialization of technology (Bozeman et al., 1999) and when addressing intellectual property and the government becoming a partner in developing the technology. These are not new topics and are found in numerous federal reports such as the reports mentioned previously, the 2019 CRS report (R45521), and the 2017 General Accountability Office report (Sullivan, 2017).

Other Transactions' appeal is that these agreements are not procurement contracts and are an alternative contracting method consistently highlighted to help alleviate these types of concerns. In addition, other transactions are designed to promote shared interests between the public and private sectors through several Congressional innovation policies. For example, Congressional actions that expanded definitions of nontraditional defense contractors to include small businesses and increased the scope of prototype projects (PL 113-291, 2014) influenced how the Department of Defense implements these policies. In addition, these policies impact alternative contracting methods (i.e., other transactions) to promote innovative technologies and products. Chapter 5 of this book, will look specifically at the policy passed in 2015 in which Congress made other transactions authority permanent by codifying the law at 10 USC 4022 (PL 114-92, 2015). By using the cooperative model as a lens to examine how innovation policies such as 10 USC 4022 influence Department of Defense's alternate contracting activities, the case study in Chapter 5 will also explore how this policy can promote innovative technologies and products using interrupted time series analysis.

Overview of Federal Contract Laws and Regulations

The procurement contract is a top-down regulated approach to procure goods and services from the private sector. These regulations are adopted by any agency that desires to do business with the private sector. As mentioned previously, other transactions are not subject to the same regulations and rely heavily on the contracting officers' discretion to enforce and shape the policy governing these types of transactions. The procurement contract method of the acquisition process (Figure 2.1) begins with the government determining a need. Per the Federal Acquisition Regulations (FAR) Part 7, before the government acquires a product or service, it must conduct market research to determine which solution is most suitable to meet the specific need. Next, the contracting officer will use the FAR and any applicable supplement such as the Defense Federal Acquisition Regulation Supplement (DFARS) to procure the product or service based on the type of product they seek to obtain. For example, "Department of Defense may use commercial item acquisition

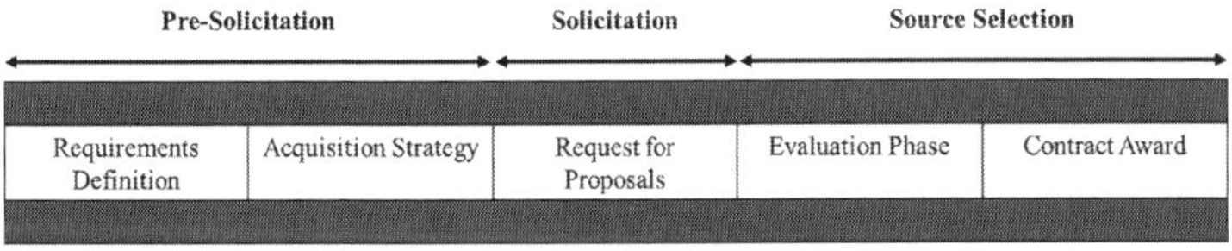

Pre-Solicitation		Solicitation	Source Selection	
Requirements Definition	Acquisition Strategy	Request for Proposals	Evaluation Phase	Contract Award

Figure 2.1 Procurement Contract Method of the Acquisition Process

Note. This procurement contract method in the acquisition process can be broken into a three-phase approach: pre-solicitation, solicitation, and source selection phase.

procedures under FAR Part 12 to procure commercially available products and negotiated contract procedures under FAR Part 15 for military-unique products" (Sullivan, 2017, p. 4). These sections of the FAR provide processes, guidelines, and applicable exceptions for contracting officers to follow in their decision-making process.

As identified in GAO-17–644, the current acquisition process is lengthy and complicated, resulting in a barrier for private sector companies to do business with the public sector. To put this into perspective, "in January 2017, the Army Contracting Command established standard contracting timelines that ranged from 55 days (about two months) for contracts valued less than $25,000 to 700 days (about 24 months) for contracts valued over $1 billion" (Sullivan, 2017, p. 13). As a comparison, "data collected by the US Air Force show that in the fiscal year 2016, it took an average of nearly 13 months from the time a request for proposal (RFP) was issued until an award decision was made for 52 sole-source contracts valued between $50 million and $500 million" (Sullivan, 2017, p. 13). This length of time can be a significant constraint for companies seeking a vehicle to complete R&D and prototyping efforts.

The contract terms and conditions can also be a significant deterrent. Per the FAR, DFARS, and the government agencies' policies, the standard terms and conditions for federal contracts are unique to the government agency. An example is a requirement in FAR Part 30 for companies to establish a government-unique cost accounting system when awarded cost-type contracts. The cost accounting system allows companies to disclose their costs in a specific manner to ensure consistency and accountability. An example provided in GAO-17–644, one company stated it took them "at least 15–18 months and cost millions to establish a government-unique cost accounting system" (Sullivan, 2017, p. 17). Government officials must track requirements from both a compliance and a liability standpoint. This increases the private sector's cost and deters businesses from working with the government.

One of the critiques of the top-down model is that it is weak when there is "no dominant policy, but rather a multitude of governmental directives and actors, none of them preeminent" (Sabatier, 1986, p. 30). The necessary discretion needed to utilize other transactions makes them an attractive alternate contracting method for the government agency among the challenges facing innovation policy. The 2016 Annual Industrial Capabilities Report to Congress stated that Congress believed other transactions are attractive for companies that do not generally engage in contracting with the government due to entry barriers and the "one size fits all" regulations governing defense procurements. The 2016 report also stated that other transactions could support the Department of Defense's efforts to attract new technological innovation offerings, specifically from Silicon Valley startups and small commercial firms.

Procurement contracts must follow the appropriate federal procurement laws and regulations. Conversely, other transactions are legally binding contracts exempt from federal procurement laws and regulations (e.g., Competition in Contracting Act and Federal Acquisition Regulations (FAR)). Most federal government acquisition statutes and regulations do not apply to the OT authority, giving contracting officials more flexibility and freedom. For example, private sector companies are not subject to following the government accounting rules (also known as Cost and Accounting Standards (CAS) as prescribed by FAR part 30). In other words, the contract resembles a private sector (B2B) contract as the terms and conditions are negotiable. The *Department of Defense Other Transaction Guide for Prototype Projects* (2018) states, "[T]his acquisition authority, when used appropriately, is a vital tool that will help the Department to lower barriers to attract nontraditional defense contractors and increase access to commercial solutions for defense requirements" (p. i). Other transactions intend to provide benefits to the Department of Defense such as attracting nontraditional contractors (as defined by 48 CFR § 212.001), establishing a network for resources to develop and obtain innovative technologies, and provide an instrument for the Department of Defense to influence technology and innovation as it did in the past.

Other transactions provide a bottom-up emphasis, giving significant decision-making power to public administrators. These public administrators have considerable influence in shaping and enacting policy on the ground, especially when it lacks clear direction on its implementation (Hill, 2003). However, implementation difficulties may occur in cases where "implementing agents know multiple ways to implement a policy and must choose among them" (Hill, 2003, p. 5). One primary reason is that other transactions are not subject to the same regulations as procurement contracts and rely heavily on contracting professionals' discretion to enforce and shape the policy governing these types of transactions. The authority provided by 10 USC 4022 gives contracting professionals decision-making power versus the regulation-driven procurement contract method. This can be thought of as a bottom-up approach to contracting, which is an alternative way to examine the public procurement model. Scholars such as Hjern and Hull (1982) who support the bottom-up approach to policy argue that examining only the perspective of "central" decision-makers neglects other actors that also play integral roles in policy implementation (Sabatier, 1986). In a similar manner, one of the top-down model critiques is that the model is weak when there is "no dominant policy, but rather a multitude of governmental directives and actors, none of the preeminent" (Sabatier, 1986, p. 30).

Tensions Between Innovation and Procurement Policy

Governments, donors, and other practitioners in the policy development community are keen to determine program effectiveness with broad goals such as increasing innovative technologies and products in the US. As mentioned in the previous chapter, *innovation policy* is a relatively new term when discussing the policymaking agenda (Edler & Fagerberg, 2017). *Innovation policy* has numerous definitions because "much of what is called innovation policy today may previously have gone under labels such as industrial policy, science policy, research policy, or technology policy" (Edler & Fagerberg, 2017, p. 5). For example, Edler and Fagerberg (2017) note that to determine the origins of the terms, one must decide if the phrase uses the qualifier "innovation," or the impact of the policy is innovation (Edler & Fagerberg, 2017). For purposes of this book, innovation policy is used to mean any policy that promotes innovation. This definition includes policies that directly support innovation using funding mechanisms such as grants and contracts or indirect support such as incentive tax programs for the private sector matching the private firm's expenditure with public funding (Vonortas, Rouge, & Aridi, 2014). This distinction is necessary because it highlights the need to assess how an intervention affects outcomes.

Another important connection is how innovation policy is implemented, specifically with the influence of public versus private sector. The challenges presented in GAO's report GAO-17–644 demonstrate the problems the Department of Defense has encountered, preventing it from attracting innovative contractors to collaborate in the development of technologies for national defense. The challenges highlighted previously, and the need for public and private sectors to work together, demonstrate how neither the market model nor the mission model is efficient in promoting innovation. Rather, these illustrate the assumptions of the cooperative model. The government can have an active part in developing innovative technologies and products.

An interest in policy and innovation led to inquiries into studies examining the relationship between policy and innovation. "From the 1970s onward, Douglas North, Robert Thomas, Nathan Rosenberg, and other economic historians argued that innovation was aided by specific government institutions and policies" (Taylor, 2016, p. 307). Policy implementation is the process of carrying out a government decision (Berman, 1978) by transforming a policy idea into an action intended to alleviate a social problem (Lester & Goggin, 1998). These actions may result in programs, procedures, or regulations (DeGroff & Cargo, 2009). One way that policymakers can use policy to promote innovation is by using public procurement to stimulate innovative activity, especially among small businesses (Vonortas et al., 2014, p. 16). Examining the outcomes of other transactions can provide insight into how policy can promote the development of innovative technologies and products.

The connection between innovation and public-private partnerships is inherent (Roumboutsos & Saussier, 2014) because the relationship calls for the two sectors to "jointly develop products and services and share risks, costs and resources which are connected with these products" (van Ham & Koppenjan, 2002, p. 598). As a result, positive outcomes and some type of efficiency gains through the private sector's involvement in providing goods and services should occur. Roumboutsos and Saussier (2014) pointed out that public-private partnerships allow for sharing resources, knowledge, and risks to support innovation in ways traditional contracting activities cannot (Roumboutsos & Saussier, 2014). Scholars have echoed the importance of the alignment of goals and values in public-private partnerships to help ensure that the public goals are met using this type of arrangement (Clark, Heilman, & Johnson, 1997; Kettl, 1993; Lombard & Morris, 2012; Savas & Savas, 2000).

This case study in Chapter 5 seeks to fill this gap by providing a documented methodology to identify and discuss whether innovation policies, such as 10 USC 4022 (PL 114-92, 2015), influence the Department of Defense's alternative contracting activities to promote the development of innovative technologies and products. By examining the innovation policies' influence on alternative contracting activities, this research aims to provide a new perspective through the lens of public administration and policy to encourage more research from diverse fields to promote policy innovation. This pilot study provides a method to look at the implementation of the policy, examines the policy with the most recent changes, examines how these policy changes affect the award rate of Department of Defense other transactions, and understands its effect on alternate contracting methods.

References

Berman, P. (1978). The study of macro- and micro-implementation. *Public Policy*, *26*(2), 157–184. PMID: 10308532.

Borrás, S., & Edquist, C. (2013). The choice of innovation policy instruments. *Technological Forecasting and Social Change*, *80*(8), 1513–1522.

Bozeman, B., Crow, M., & Tucker, C. (1999). *Federal laboratories and defense policy in the US national innovation system*. Paper presented at the Summer Conference on National Innovation Systems, Rebild.

Clark, C., Heilman, J. G., & Johnson, G. W. (1997). Privatization of wastewater treatment plants: Lessons from changing corporate strategies. *Public Works Management & Policy*, *2*(2), 140–147. doi:10.1177/1087724x9700200204

DeGroff, A., & Cargo, M. (2009). Policy implementation: Implications for evaluation. *New Directions for Evaluation*, 47–60. doi:10.1002/ev.313

Edler, J., & Fagerberg, J. (2017). Innovation policy: What, why, and how. *Oxford Review of Economic Policy*, *33*(1), 2–23. doi:10.1093/oxrep/grx001

Hill, H. C. (2003). Understanding implementation: Street-level bureaucrats' resources for reform. *Journal of Public Administration Research and Theory*, *13*(3), 265–282.

Hjern, B., & Hull, C. (1982). Implementation research as empirical constitutionalism. *European Journal of Political Research*, *10*(2), 105–115.

Kettl, D. F. (1993). *Sharing power public governance and private markets*. Washington, DC: The Brookings Institution.

Lester, J. P., & Goggin, M. L. (1998). Back to the future: The rediscovery of implementation studies. *Policy currents*, *8*(3), 1–9.

Lombard, J. R., & Morris, J. C. (2012). Using privatization theory to analyze economic development projects. *Public Performance & Management Review, 35*(4), 643–659. doi:10.2753/PMR1530-9576350404

Mazmanian, D. A., & Sabatier, P. A. (1981). *Effective policy implementation.* Free Press.

Mazzucato, M. (2015). *The entrepreneurial state: Debunking public vs. private sector myths* (Vol. 1). Anthem Press.

Roumboutsos, A., & Saussier, S. (2014). Public-private partnerships and investments in innovation: The influence of the contractual arrangement. *Construction Management & Economics, 32*(4), 349–361. doi:10.1080/01446193.2014.895849

Sabatier, P. A. (1986). Top-down and bottom-up approaches to implementation research: A critical analysis and suggested synthesis. *Journal of Public Policy, 6*(1), 21–48.

Savas, E. S., & Savas, E. S. (2000). Privatization and public-private partnerships. Los Angeles, CA: SAGE Publications.

Sullivan, M. J. (2017). *Military acquisitions: Department of Defense is taking steps to address challenges faced by certain companies* (Report no. GAO-17-644). United States Government Accountability Office.

Taylor, M. Z. (2016). *The politics of innovation: Why some countries are better than others at science and technology.* Oxford University Press.

van Ham, J. C., & Koppenjan, J. F. M. (2002). Port expansion and public-private partnership: The case of Rotterdam. *WIT Transactions on the Built Environment, 62*.

Vonortas, N. S., Rouge, P. C., & Aridi, A. (2014). *Innovation policy: A practical introduction.* Springer.

3 Traditional Contract Pathways and Strategies for R&D

Public Procurement and the Contracting Process

Public procurement has become essential to support the demand for valuable and cost-saving services. The goals of public procurement include cost, quality, timeliness, risk management, accomplishing social and economic objectives, and maximizing competition while maintaining integrity and transparency (Pitzer & Thai, 2009). The procurement department of any government agency must ensure proper leadership, management, and capability to meet these goals. This includes training on federal, state, and local regulations and ensuring compliance across the board. Procurement departments must confirm that these regulations are efficient and effective. Due process requires time, effort, and cost. Thus, there is a trade-off between transparency, timeliness, and efficiency (Pitzer & Thai, 2009). Aside from internal forces, procurement departments face external goals and challenges, including social, economic, and political.

The contract development stages of public procurement depend on the strength of the pre-planning and the strategic planning phases. Failure to communicate in the pre-proposal stage can result in negative consequences that hinder the competitive environment (Curry, 2010). If the planning process is lacking, then the solicitation can contain features that encourage conflict of interest. This can lead to the proposal evaluation, resulting in selecting a contractor that needs to meet the requirements (Curry, 2010). Another consequence of poor communication is the perception of unequal treatment, ensuing possible protests. Legitimate protests can hurt a department's reputation, leading to pressures that may engender ethical violations, procurement fraud, and conflicts of interest.

Contract performance is monitored for quality, cost, and schedule. One way to minimize risk is to ensure that the individuals handling the procurement process are appropriately trained in the regulations governing their actions. These regulations apply to both public and private organizations in public procurement. Public officials are encouraged to consult the enabling legislative and administrative laws for best practices (Pitzer & Thai, 2009).

22 Traditional Contract Pathways and Strategies for R&D

The procurement officials can consult their legal and contracts department to understand the requirements of the laws and regulations that are applicable. The contracts department has specialized expertise in interpreting applicable regulations.

The model of a contracting arrangement in privatization provided by Savas (1987) displays a relationship between three entities (p. 68). Figure 3.1 illustrates the arrangement, depicting the Government (G), the Private Firm (F), the Consumer (C), and the flow of authorization between the parties. The figure demonstrates where the government authorizes (solid line) and pays a private firm (dotted dashed line) to deliver (dashed line) a good or service to the consumer (Savas, 1987, p. 68).

The contract arrangement displayed in Figure 3.1 illustrates the relationship between the government and the private firm. Savas (1987) highlights that government contracts can be with other governments, private firms, and non-profit organizations (p. 68). The term *public-private partnerships* describes the relationship created when public organizations use private organizations to deliver goods and services (Custos & Reitz, 2010). Savas (1987) examines goods and services constructed on two properties: exclusion and consumption (p. 35). The degree of these two properties classifies the goods and services into four categories: private goods, toll goods, common goods, and collective goods (1987, p. 56). Understanding the type of goods leads to understanding the service models that help obtain these goods. These models include, but are not limited to, government service, government vending, contracts, franchises, and grants (1987, p. 91). While these

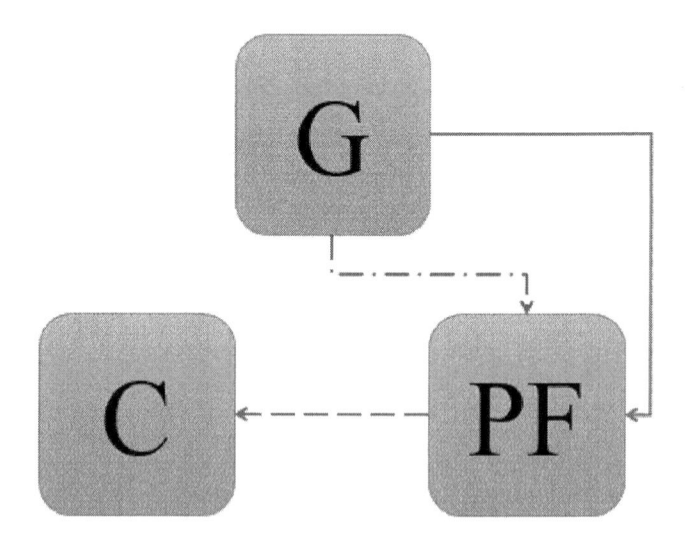

Figure 3.1 Contracting Arrangement of Privatization

models are related to the good or service they provide, public-private partnerships are also closely related to the regulatory actions that govern their activity. These regulations define the relationship, the vehicle model, and other appropriate specifications.

Private organizations also use contract arrangements in their business operations. Savas (1987) provided a model for this relationship labeling it as the market arrangement (p. 79). This arrangement is defined by the exclusion of government in the transaction (p. 79). Figure 3.2 displays the market arrangement depicting the Government (G), the Private Firm (F), the Consumer (C), and the flow of authorization between the parties. The figure illustrates where the consumer authorizes (solid line) and pays a private firm (dotted dashed line) to deliver (dashed line) a good or service to the consumer (Savas, 1987, p. 80).

In comparison to Figure 3.1, Figure 3.2 shares some similarities in the entity interaction. The entity that authorizes the transaction also pays for the delivery of the goods or services to the end consumer. Another similarity in these two models is that the private firm delivers the goods or services to the end customer in both arrangements. An arrangement that Savas did not account for is the situation in which the private firm in Figure 3.1 decides to participate in a separate market arrangement, Figure 3.2, to provide the good or service initiated by the contract arrangement in Figure 3.1. In this situation, the roles are no longer clearly defined.

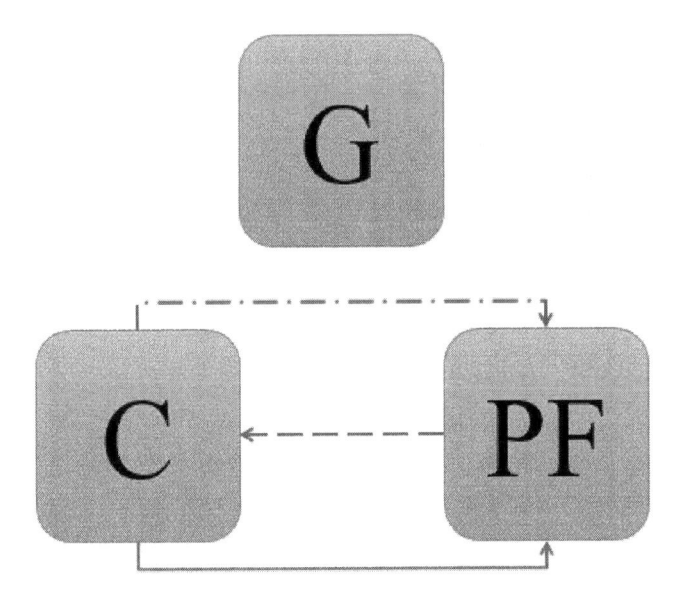

Figure 3.2 Market Arrangement

24 Traditional Contract Pathways and Strategies for R&D

Principal-Agent theory can provide clarity to this situational relationship. Principal-Agent theory is vital in understanding public-private partnerships and their association with contracting (Neill & Morris, 2012). The approach of Principal-Agent theory is a framework of relationships between the government and the private sector (Kettl, 1993, pp. 22–25). The government, acting as the principal, enlists an agent, a company in the private sector, to perform a specific scope of work. The simplest, and most common, application of this theory is the interaction between one principal and one agent that can avert risk from the principal (Lane, 2013).

Neill and Morris (2012) suggest that applying this theory in a simplistic manner is not practical because in real-life applications, it tends to be more complex – the relationship can involve multiple principals and multiple agents (p. 632). Figure 3.3 provides an illustration of the complex system that these relationships, as provided by the framework of Neill and Morris (2012). Examining Figure 3.3, with the application of the contracting arrangement depicted in Figure 3.1, complexity that public-private partnerships experience becomes evident. The basic goal of the relationship is to encourage the agent to perform as the principal desires. Lane (2013) makes the point that contracts have two basic principles to satisfy: the price and the incentive (p. 86). Given perfect information, the process is simple and can result in the ultimate solution (Lane, 2013). However, as depicted in Figure 3.3, the process is not simple. The complexities of the principal-agent model and public-private partnerships result in exposure to a critical problem known as *information asymmetry* (Lane, 2013; Neill & Morris, 2012; Kettl, 1993).

Information asymmetry results from imperfect information limiting competition (Kettl, 1993). In contract theory and economics, information asymmetry

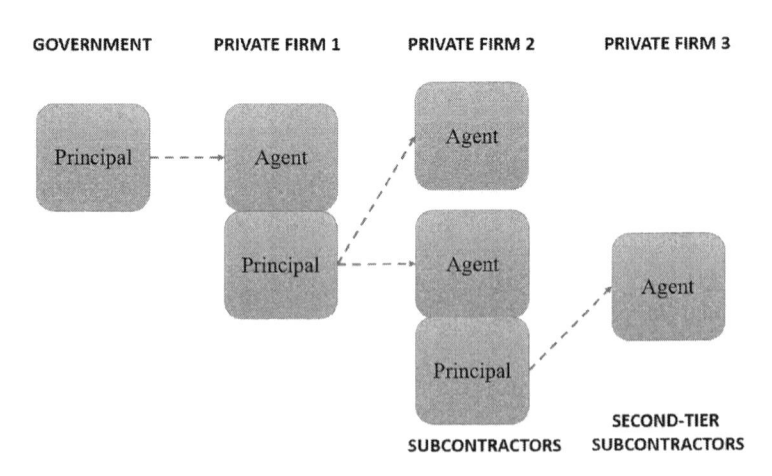

Figure 3.3 Complex Principal-Agent Relationship in Contracting

deals with the study of decisions in transactions where one party has either more or better information than the other party in the relationship. By limiting competition, information asymmetry hinders the one critical benefit of leveraging public-private partnerships. As mentioned before, competition is leveraged from the private sector to overcome the inefficiencies in the government and provides a driving force for innovative solutions. The argument of limited information was expanded by Simon (1947), stating that it was "impossible for behavior of a single, isolated individual to reach any high degree of rationality" due to the limitation of information (p. 92). The consequence of this conflict is that it leads to subpar decisions and poorly aligned goals (Lane, 2013). The application of this theory provides context for the relationship and the impact decision-making can have on a public-private relationship.

During the formation of the principal-agent relationship, the public organizations must establish goals to incentivize the desired performance from the agent. However, the goals provided by the principal are generated to favor the principal's interest (Neill & Morris, 2012, p. 633). Understanding that goals should be aligned for each party in the relationship is essential. For example, suppose goals are misaligned and in favor of the principal. In that case, the principal may indirectly be encouraging the agent to engage in shirking to maximize the agent's outcome in the contractual arrangement (Neill & Morris, 2012; Kettl, 1993). Thus, to minimize shirking, the principal monitors behaviors, further contributing to asymmetry in the relationship.

To strike the appropriate balance and create suitable incentives, the principal must understand how forming goals will contribute to the overall decision-making process. As Kettl (1993) highlights, the goals in a contract arrangement are crucial to establishing the boundaries of the agreement (p. 25). Goals help define responsibilities and work efforts. If information is fragmented, decision-makers cannot choose a satisfactory solution and poor performance on a contract.

Traditional Methods the Government Contracts Out

The Federal Acquisition Regulations (FAR), which has its roots in the Armed Services Procurement Regulation established in 1947, are the principal set of rules regarding government procurement in the US. Effective 1 April 1984, the FAR is codified in Chapter 1 of Title 48 of the Code of Federal Regulations (CFR) and divided into 53 parts, each dealing with a separate aspect of the public procurement process. The first six parts cover general government acquisition matters, followed by another six that cover acquisition planning aspects; the rest of the FAR covers topics such as simplified acquisition, thresholds, large dollar value buys, labor laws, contract administration, applicable clauses, and forms. Other highlights of the FAR include parts covering information related to small businesses (FAR Part 19, Small Business Program) and standard terms and conditions in a government contract (FAR Part 52, Solicitation Provisions and Contract Clauses). The purpose of the clauses is to provide a set of consistent policy procedures within the federal acquisition process to keep the government contracting ecosystem flowing smoothly. Government acquisition professionals must follow the FAR, internal policies, and their agency supplements, such as the Defense Federal Acquisition Regulation Supplement (DFARS), the General Services Acquisition Regulation Supplement (GSARS), and the National Aeronautics Space Administration FAR Supplement (NFS). It is important to note that these clauses, and the requirements within, change based on the type of company the government wants to engage. This chapter looks at two FAR parts applicable to innovation, small business programs, and statute applicable to innovation.

Federal Acquisition Regulations Part 12 (48 CFR Part 12)

The Federal Acquisition Streamlining Act of 1994 (FASA), Public Law No: 103-355, was a piece of acquisition reform legislation to streamline the US federal government's acquisition system and to change how it performs its contracting functions dramatically. The legislation had several objectives focused on increasing the government's dependence on commercial goods and services and streamlining the procurement process to improve access by small businesses to government contracting opportunities. A key provision in FASA is the strongly stated preference for buying commercial "off-the-shelf" items rather than purchasing through the detailed bidding process for government-unique items. The legislation provided broad definitions for commercial products and services to incentivize private firms to do business with the government. It eliminated numerous statutory requirements for purchases of such items to simplify the process.

FAR Part 12 implements the US federal government's preference for procuring commercial products and services contained in 41 USC 1906, 1907, 3307, and 10 USC 3451–3453 in Title VIII of the FASA. The concept of FAR Part 12 is to take advantage of products and services already available to the general public and establish procurement policies similar to how the private sector operates. For example, FAR Part 12 requires that federal agencies conduct market research to determine whether commercial products and services are already available that can meet the government's requirements. If those commercial products and services are already available, then the government should purchase the commercial products and services. This requirement is not just limited to the government – in the contract with private sector, the government agency requires prime contractors and subcontractors at all tiers to incorporate commercial items as components of items supplied to the agency.[1] In many instances, the requirement is flowed down from one party of the agreement to another to ensure that each party has the same responsibility to prioritize commercial products and services.

Federal Acquisition Regulations Part 35 (48 CFR Part 35)

FAR Part 35, titled Research and Development Contracting, first appeared in the Federal Register in 1983 and has been part of the FAR since its inception. The purpose of this FAR part is to establish policies and procedures for research and development (R&D) contracts.[2] One notable attribute in FAR Part 35 is a solicitation method called Broad Agency Announcement (BAA), FAR part 35.016, which first appeared in the Federal Register on 20 July 1988[3] and was codified as 48 CFR 35.016. Agencies may use BAAs to fulfill their basic and applied research requirements and to increase their understanding of certain areas of interest. The BAA technique is unique because it is used only when meaningful proposals with varying technical/scientific approaches can be reasonably anticipated. The government can use any preferred contract or agreement for the type of effort required. This can include a grant, cooperative agreement, or other transaction (covered in the next chapter). The contract or agreement of preference must be clearly articulated in the solicitation.

Interested companies can submit white papers to the government based on the instructions of the BAA solicitation. The white papers are then reviewed by a panel against the criteria stated in the solicitation. The government will contact all offerors submitting white papers, either informing them that the effort proposed is not of interest to the government or requesting a formal cost and technical proposal by a specified date. To ensure all technical proposals receive proper consideration, each proposal must follow the format provided by the government. The government may award some, all, or none of the proposals. If an award is being made, typically, more than one award is issued on the basis of the quality of the proposals and availability of funding.

Small Business Innovation Research Program

The Small Business Innovation Development Act of 1982 (PL 97-219) established the Small Business Innovation Research (SBIR) program to strengthen the role of innovative small business concerns in federally funded research and development. This program aims to stimulate technological innovation, prioritize small businesses to meet US research and development needs, and increase private sector commercialization of innovation, funded by the US government. The SBIR program is codified at § 9 of the Act, 15 USC 638. Federal agencies with an extramural research and development budget of more than $100 million must participate in the SBIR program and obligate at least 3.2% of the respective budget to small businesses under the SBIR program. To put this in perspective, in fiscal year 2019 (1 October 2018–30 September 2019), the federal government made over 179,000 awards totaling more than $54.3 billion. There are currently 11 federal agencies participating in the SBIR program.

The SBIR/STTR Reauthorization Act (PL 117-183) was passed on 30 September 2022 and does not expire until 2025. This reauthorization came with some updates to the current Small Business Administration Policy Directive and included six pilot programs that will expire by the end of the extension. These reauthorization and other legislative updates also focused on reporting on the awards through these two programs. It included additional directions for government agencies to gather important information from the small businesses submitting their proposals.

Small Business Technology Transfer Program

Modeled after the SBIR program, the Small Business Technology Transfer (STTR) program was introduced as a pilot program by the Small Business Technology Transfer Act of 1992 (PL 102-564, Title II). The STTR program's statutory purpose is to stimulate a partnership between innovative small businesses and non-profit research institutions, which supports the goals for the commercialization of innovative technologies. The STTR program is also codified at § 9 of the Act, 15 USC 638. Unlike the SBIR program, the STTR program has a higher budget, requiring federal agencies with extramural research and development budgets of more than $1 billion to participate in the STTR program and obligate at least 0.45% of the respective budget. Currently, five federal agencies are participating in the STTR program.

The US Department of Defense (DoD) is the biggest funder for both programs, which means that any modifications to the SBIR program, the STTR program, and other similar innovation programs have been passed in the National Defense Authorization Act (NDAA). Many policy analysts track the annual release of the NDAA and its drafts for any trends to prepare for each new fiscal year, which runs from October to September every year. A key attribute of these two programs is the three-phase approach to contracting (Kuchina-Musina, 2021). These three phases are as follows.

- Phase I: The purpose of Phase I is to explore the technical merit and feasibility of an idea or technology and determine the commercial potential of the proposed effort. In addition, during this phase, the small business's performance quality is evaluated before advancing to Phase II. Phase I contracts are no more than six months for a value of $50,000–$250,000. Funding must be through the SBIR/STTR program.
- Phase II: Phase II aims to continue the R&D efforts initiated in Phase I. The funding in Phase II is based on the performance during Phase I and the proposed efforts' technical merit and commercial potential of the Phase II project. Typically, only Phase I awardees are eligible for Phase II awards. Phase II contracts are for 24 months or less, are funded by the SBIR/STTR program, and have a value between $750,000 and $1,500,000.

- Phase III: The efforts during Phase III derive from, extend, or complete efforts under prior SBIR Phases and enable a business to pursue commercialization. Phase III work may be for products (including test and evaluation), production contracts, or R&D activities. The SBIR program does not fund Phase III awards. Phase III contract value and duration do not have a limit.

The SBIR/STTR program continues to evolve and remains the primary source of early funding to thousands of highly successful small businesses. Many of these awardees leverage opportunities in the program to become large businesses gradually, and some have become industry leaders. The recent economic impact studies developed by the US Air Force, Navy, DoD, and National Cancer Institute demonstrate that the program generates one of the highest returns on research and development dollars for the federal government.

Federal Prize Competitions

The COMPETES Reauthorization Act of 2010 (PL 111-358) was the first legislation to authorize federal agencies to carry out prize competitions to promote innovation to advance the federal agencies' mission. The American Innovation and Competitiveness Act of 2017 (PL 114-329) made several technical and clarifying amendments to this broad prize competition authority, including language authorizing the head of a federal agency to request and accept funds from other federal agencies, state and local government, or private sector to design and administer prize competition or for the cash prize. Over the years, Congress has provided some federal agencies with additional explicit authority to conduct prize competitions. There are six federal agencies with specific prize competition authorities, including the Department of Defense, the Department of Energy, the National Aeronautics and Space Administration, the Department of Health and Human Services, the Department of Transportation, and the Department of Commerce.

Due to Congress's increased interest in using prize competitions, federal agencies have invested in developing more in-house expertise in the design and administration of prize competitions. The Congressional Research Service report R45271 points out the following:

> For example, in FY2017 and FY2018, eight federal agencies had department-wide policy and guidance on the use of prize competitions; five agencies had dedicated, full-time prize competition personnel; sixteen agencies had distributed networks or communities of prize and project managers with prize competition expertise within the agency; and five agencies were providing centralized training and design support to agency staff (p. 10).

Procurement for Experimental Purposes

On 20 November 1993, the National Defense Authorization Act for Fiscal Year 1994 (PL 103-160) established the Procurement for Experimental Purposes, which authorized the Department of Defense to procure the necessary number of items for experimentation and technical and operational evaluation or to maintain a capability. This authority currently allows procurement in nine specific areas: ordinance, signal, chemical activity, transportation, energy, medical, spaceflight, aeronautical supplies, and telecommunications. A notable attribute to these procurements is that the FAR and DFARS are not applicable to them, meaning that the required formal competitive procedures also do not apply. The Department of Defense can also leverage its other transaction authorities to execute research or prototype other transaction agreements for the nine items allowed under this statute.

This authority has positives and negatives, just like any public procurement method does. Some positives include a flexible and fast contract vehicle to acquire products outside the US, which can be "stacked" with other statutory authorities if long-term, critical thinking is applied to acquisition strategy across acquisition phases. On the other hand, some of the negatives of this authority are that use is still relatively unknown; lack of guidance and established precedent increases the risk to the government. It requires highly experienced and empowered staff, which can deter the Department of Defense from using this method to procure innovative technologies.

Summary

The topic of contract development and management has been gaining attention due to the pressures of budget cuts and the demand for increased efficiencies (Jurisch, Ikas, Wolf, & Krcmar, 2013). The private sector seeks efficiency and depends on competition to thrive, while the government is driven by budget requirements and resource allocation (Kettl, 1993; Jurisch et al., 2013; Berrios, 2006). These distinct differences play a vital role in understanding public-private partnerships and each organization's process to decide if they will contract out. The current measure of effectiveness across all agencies is reporting the number of contracts issued and which phase, the amount of money obligated, and other factors that may vary from agency to agency. These metrics must be reported to Congress through reporting tools such as Federal Procurement Data System – Next Generation (FPDS-NG). These metrics are later compiled into an annual public information report issued by each federal agency or an oversight agency like the General Accountability Office (GAO). These reports can measure many factors and the variance between agencies and use cases. Some of the variance is the product of differences at the agency enterprise level, and others originate from different approaches to running the program.

Policymakers are committed to evaluating these differences and encouraging agencies to adopt best practices. Data from this report are crucial to assessments of innovation development, the use of these contract pathways, and the obligation of federal funding.

Notes

1. See also 41 USC 3307.
2. R&D integral to the acquisition of major systems is covered in FAR Part 34. Independent research and development (IR&D) is covered at 31.205–18.
3. https://archives.federalregister.gov/issue_slice/1988/7/20/27449-27468.pdf#page=19

References

Berrios, R. (2006). Government contracts and contractor behavior. *Journal of Business Ethics, 63*(?), 19–130. Retrieved from www.jstor.org.proxy.lib.odu.edu/stable/25123695

Curry, W. S. (2010). *Government contracting: Promises and perils*. Boca Raton, FL: Taylor and Francis Group.

Custos, D., & Reitz, J. (2010). Public-private partnerships. *The American Journal of Comparative Law, 58*, 555–584. Retrieved from www.jstor.org.proxy.lib.odu.edu/stable/20744554

Gallo, M. E. (2020). *Federal prize competitions* (Report No. R45271). Washington, DC: Congressional Research Service.

Jurisch, M., Ikas, C., Wolf, P., & Krcmar, H. (2013). Key differences of private and public sector business process change. *E-service Journal, 9*(1), 3–27. doi:10.2979/eservicej.9.1.3

Kettl, D. F. (1993). *Sharing power: Public governance and private markets*. Washington, DC: Brookings.

Kuchina-Musina, D. (2021). An imposter among us? – SBIR phase 3. *Public Spend Forum Editorial Staff*. Retrieved from www.publicspendforum.net/blogs/psfeditorial/2021/02/17/an-imposter-among-us-sbir-phase-3/

Lane, J. (2013). The principal-agent approach to politics: Policy implementation and public policy making. *Open Journal of Political Science, 3*(2), 85–89. http://dx.doi.org/10.4236/ojps.2013.32012.

Neill, K. A., & Morris, J. C. (2012). A tangled web of principals and agents: Examining the deepwater horizon oil spill through a principal-agent lens. *Politics & Policy, 40*(4), 629–656. doi:10.1111/j.1747-1346.2012.00371.x

Pitzer, J. P., & Thai, K. V. (2009). *Introduction to public procurement*. Herndon, VA: National Institute of Governmental Purchasing.

Savas, E. S. (1987). *Privatization: The key to better government*. Chatham, NJ: Chatham House.

Simon, H. A. (1947). *Administrative behavior: A study of decision-making processes in administrative organization*. New York: Ronald Press.

4 Alternative Contract Pathways for R&D

Technology Transfer and Alternative Contract Pathways

A growing challenge in public procurement organizations is pursuing creativity, innovation, and change-oriented behaviors that challenge the status quo. The difficulty in fostering these initiatives in public organizations is attributed to Weber's theory of bureaucracy (Isett, Glied, Sparer, & Brown, 2013). This formally structured organization depends on the systematic process and organized hierarchies to maintain order, maximize efficiency, and eliminate favoritism (Weber, 1947). However, the dynamic and competitive environment makes it harder for public organizations to ignore the need for change (Lutz Allen, Smith, & Da Silva, 2013; Vigoda-Gadot & Beeri, 2012). Factors such as demands from the new labor pool, the decline of resources, and the need for talent retention (Paarlberg & Lavigna, 2010) pressure public organizations to find innovative approaches to archaic procedures and work practices (Vigoda-Gadot & Beeri, 2012). So, how do organizations foster change and promote creativity and innovation?

Organizational creativity has been defined as "the creation of a valuable, useful new product, service, idea, procedure, or process by individuals working together in a complex social system" (Woodman, Sawyer, & Griffin, 1993, p. 293). Transformational leadership promotes creativity by encouraging employees to question the status quo (Jong & Hartog, 2007; Lutz Allen et al., 2013). Leaders of public procurement organizations of today (or of tomorrow) do not necessarily need to be transformational leaders, but if they are, their impact on organizational change must support the public interest and align with the agency's mission, vision, and goals. In addition, public organizations must be careful when hiring leaders to ensure that the individual goals support the public interest and align with the agency's mission, vision, and goals. The consequences of hiring a transformational leader and not aligning their creative and innovative characteristics include cost and value implications on the procurement department and projects.

In innovation, technology transfer is the crucial process through which technology originating from federal labs, universities, or research institutions is transferred to the private sector for potential commercialization. The

DOI: 10.4324/9781003398455-4

US government invests heavily in research and development, with an annual budget exceeding $100 billion, resulting in a continuous flow of innovative inventions and technologies. Technology transfer facilitates the transformation of these inventions into commercially viable products or services with the help of industry partners who further develop, scale up, and bring them to the market.

Starting with the Stevenson-Wydler Technology Innovation Act of 1980 (PL 96-517), Congress has created a series of laws to promote and encourage technology transfer. These laws encourage the private sector to harness federally funded technology through various mechanisms, including Cooperative Research and Development Agreements (CRADAs), establishing startup companies, patent license agreements, educational partnerships, and collaborations with state and local governments. This framework bridges the gap between scientific research and practical applications, fostering innovation and economic.

Stevenson-Wydler Technology Innovation Act of 1980

The Stevenson-Wydler Technology Innovation Act of 1980 (PL 96-480) (94 Stat. 2311) marked a significant milestone as the initial primary US legislation pertaining to technology transfer. This law mandates that federal laboratories engage actively in technology transfer endeavors and allocate funding for such initiatives. Furthermore, it stipulates that every federal laboratory must establish an Office of Research and Technology Applications responsible for overseeing and advancing technology transfer efforts.

The Stevenson-Wydler Technology Innovation Act established the Department of Commerce Office of Industrial Technology, which oversees the establishment of a wide variety of university-affiliated centers. The centers serve four primary purposes: first, they conduct research to bolster technological and industrial innovation – fostering collaborations between industry and universities; second, they aid individuals and small businesses in nurturing technological concepts for innovation and the establishment of new ventures; third, they offer technical support to various industries, with a particular emphasis on assisting small businesses; and finally, they provide curriculum development and training opportunities in the realms of invention, entrepreneurship, and industrial innovation.

The primary focus of the Stevenson-Wydler Act was to disseminate information from the federal government to the public and to require federal laboratories to engage actively in the technology transfer process. The law requires laboratories to set apart a percentage of the laboratory budget for technology transfer activities. Additionally, this legislation established the Center for the Utilization of Federal Technology in the Department of Commerce to serve as a central clearinghouse for the collection, dissemination, and transfer of information on Federally owned or originated technology having potential application to State and local government and private industry (Jolly, 1980).

The Bayh-Dole Act

The Bayh-Dole Act, officially known as the Patent and Trademark Law Amendments Act Public Law No: 96-517 and codified in 35 USC 200–212, was enacted in 1980 with the aim of encouraging and expediting the commercialization of research outcomes funded by the federal government. This legislation empowers institutions and grant recipients, such as universities, to assume ownership of patents arising from government-funded research and subsequently license these patents to industry partners. This arrangement can yield substantial royalties for the research institution in the event of successful commercialization.

Bayh-Dole has profoundly impacted the proliferation of patenting and licensing activities among US universities. In addition, it has been a catalyst for the substantial growth in the formation of startup companies dedicated to developing and commercializing technologies funded by the federal government, operating under licenses obtained from universities.

Federal Technology Transfer Act (FTTA) of 1986

The Federal Technology Transfer Act (PL 99-502), codified in 15 USC 3710, was passed by Congress in 1986 as an amendment to the Stevenson-Wydler Act of 1980. Its primary objective is to enhance industry access to technologies from federal laboratories. This legislation led to the establishment of a nationwide network of federal laboratories, agencies, and research centers called the Federal Laboratory Consortium for Technology Transfer.

Furthermore, the Federal Technology Transfer Act empowered federal laboratories to engage in negotiations for licenses concerning patented inventions originating within the laboratory. It also authorized them to enter CRADAs (detailed later on in this chapter). CRADAs are formal, documented agreements involving one or more federal laboratories and non-federal entities, wherein the government, through its laboratories, contributes resources such as personnel, services, facilities, equipment, and intellectual property. Notably, the Federal Technology Transfer Act specifies that no funding may flow from the federal laboratories to the non-federal parties under these CRADAs.

National Technology Transfer and Advancement Act of 1995

The National Technology Transfer and Advancement Act of 1995 (PL 104-113) (110 Stat. 775) amended the Stevenson-Wydler Act, intending to make CRADAs more appealing to federal laboratories and private industry. This legislature assures US companies that they will receive adequate intellectual property rights to encourage the swift commercialization of inventions resulting from CRADAs. Moreover, the National Technology Transfer and

Advancement Act promotes the establishment of new technology standards by mandating that all federal agencies adopt cooperatively developed standards, particularly those formulated by standards-developing organizations.

Existing guidance from the Office of Management and Budget (OMB) is codified in Section 12 of the National Technology Transfer and Advancement Act, and directed federal agencies to increase their use of voluntary consensus standards and use them instead of government-developed standards whenever possible. The act assigned broad oversight responsibility to OMB and coordination responsibilities to the National Institute of Standards and Technology (NIST). In February 1998, OMB issued revised Circular A-119 (Federal Participation in the Development and Use of Voluntary Consensus Standards and in Conformity Assessment Activities) to be consistent with the terminology in the act and provide federal agencies with guidance on how to meet the act's requirements. To reduce these problems, the act (PL 104-113) and the OMB circular direct federal agencies to use voluntary consensus standards whenever possible to reduce duplication in production lines and assist U.S. competitiveness in trade. The circular describes impractical circumstances such as those that would fail to serve the agency's program needs or be inconsistent with the agency mission. The circular directs the agencies, through NIST, to provide OMB with information on their standards activities for inclusion in an annual report to the Congress.

Assistance Agreement Instruments

Assistance agreements are formal legal mechanisms crafted to transfer something of value to a recipient for the purpose of supporting or stimulating a public endeavor sanctioned by a US law (refer to 31 USC 6101(3) for details). Examples of these legal instruments encompass grants, cooperative agreements, and technology investment agreements, all employed to furnish assistance. It is crucial to distinguish assistance and assistance agreements from the concept of acquisition, as defined by regulations, which pertains to the procurement of goods and services for the direct benefit of the US government. Per 31 USC 6303, procurement contracts represent the appropriate legal instruments for acquiring such property or services.

Grants

The Federal Grant and Cooperative Agreement Act of 1977 (PL 95-224) guided government agencies in deploying Federal funds – particularly by distinguishing between contracts, cooperative agreements, and grants. This act addressed US Congressional concerns over the perceived misuse of assistance agreements to circumvent competition and procurement rules. The Federal Grant and Cooperative Agreement Act established government-wide guidelines for determining the appropriate legal instrument to fund extramural activities. The distinguishing factor between a grant and a cooperative agreement is the degree of Federal participation or involvement during work activities. An executive agency uses a grant agreement as the legal instrument when the established relationship is intended to transfer a good or service from one recipient to another for public use or benefit. Extensive involvement of the federal agency and the other recipient is not a requirement.

In 1980, the General Accountability Office published General Counsel Milton J. Socolar's interpretation of the Federal Grant and Cooperative Agreement Act of 1977 (Report B-196872-O.M.). In his interpretation, Socolar stated that the Act requires that agencies use the correct legal instrument (grant, cooperative agreement, or contract) when procuring goods or services, which is determined by analyzing the types of relationships authorized and the circumstances. Socolar states that the legal instrument (contract, grant, or cooperative agreement) that fits the arrangement as contemplated must be used once authority is found. The report outlines additional guidance that can be captured in follow-on statutes and best practices still used today.

Since 1982, Chapter 63 of the US Code, titled Using Procurement Contracts and Grant and Cooperative Agreements, provides applicable definitions and directions on using procurement contracts, grant agreements, and cooperative agreements. The statute governing Grant agreements is 31 USC 6305, titled "Utilizing Grant Agreements." According to this statute, a grant represents a formal legal document utilized to establish a relationship under the following conditions:

(a) Its primary intent is to transfer a valuable asset to the recipient to facilitate the execution of a public purpose of support or stimulation as authorized by law, as opposed to obtaining property or services for the direct benefit or use of the Department of Defense (DoD).
(b) It does not anticipate substantial involvement between the DoD and the recipient during the execution of the activity outlined in the grant.

This legislation also provided guidance on the use of cooperative agreements. These agreements are detailed below.

Cooperative Agreements

As mentioned earlier, the Federal Grant and Cooperative Agreement Act of 1977 (PL 95-224) provided guidelines for cooperative agreements. Since 1982, the statute for Cooperative Agreements is 31 USC 6305, titled Using Cooperative Agreements.[1] Pursuant to the statute, a cooperative agreement is a federal assistance instrument designed to establish a similar relationship as a grant. The significant difference between a grant and a cooperative agreement is that grants are assistance awards for which no substantial involvement is anticipated between the DoD and the recipient during the contemplated activity, whereas cooperative agreements anticipate substantial involvement.[2]

Like a grant, a federal agency should utilize a cooperative agreement as the designated legal instrument to transfer a good or service from one recipient to another for public use or benefit. However, in the instance of a cooperative agreement, extensive involvement of the federal agency and the other recipient is required and outlined in the agreement.

Technology Investment Agreement

In 1989, Congress enacted 10 USC 4021,[3] "Research projects: Transactions other than contracts and grants," to authorize the Department of Defense to use cooperative agreements and other transactions. Using this authority, the DoD developed types of cooperative agreements and other transactions to support research (called "dual-use" research) with a strong potential for both commercial and defense applications. In 1997, the DoD issued interim guidance that fused various Cooperative Agreements and other similar transactions into a single group of assistance instruments known as Technology Investment Agreements (TIAs). In 2023, the DoD added a new part of the DoD Grant and Agreement Regulations (DoDGARs) to incorporate policies and procedures for the award and administration of Technology Investment Agreements. TIAs are a relatively new class of assistance instruments to stimulate defense research projects involving for-profit firms, especially commercial firms that primarily do business in the private sector. Similar to a Cooperative Agreement, these agreements require significant federal involvement in the technical or management aspects of the project. The most significant advantage of using a TIA is that it allows DoD agreements officers to negotiate award provisions in areas that can pose obstacles to commercial firms, such as cost principles and intellectual property concerns.

Acquisition Agreement Instruments

The principal purpose of acquisition instruments is to acquire property or service for the direct benefit or use of the US government. Unlike assistance instruments, acquisition agreement instruments provide a way for the government to license or own the final product identified in the contract.

Partnership Intermediary Agreements

A Partnership Intermediary Agreement (PIA) codified in 15 USC 3715 and 10 USC 2368 represents a formal arrangement, contract, or memorandum of understanding with a non-profit partnership intermediary. This intermediary's role is to engage academia and industry on behalf of the government, with the primary aim of expediting technology transfer and licensing processes. Specifically, a PIA is described as an agency associated with a state or local government or a non-profit organization that is either wholly or partially owned, chartered by, funded to some degree by, or operated in part or whole by or on behalf of a state or local government. These intermediaries provide assistance, guidance, evaluation, or other forms of collaboration to small business enterprises and institutions of higher education seeking, or capable of, effectively utilizing technology-related support from a federal laboratory.

The authority to engage partnership intermediaries with federal labs is granted by 15 USC 3715. Federal labs, as defined under this statute, encompass any laboratory, federally funded research and development center, or center established pursuant to 15 USC 3705 (Cooperative Research Centers) or 15 USC 3707 (NSF Cooperative Research Centers). These facilities are owned, leased, or otherwise employed by a federal agency and funded by the federal government, whether operated directly by the government or by a contractor.

Educational Partnership Agreements

Section 247 of the National Defense Authorization Act for Fiscal Year 1991 (PL 101-510) introduced 10 USC 2194, titled "Education Partnerships." This legislation grants authority for the directors of defense laboratories to initiate Educational Partnership Agreements (EPAs) with educational institutions across the US. These agreements aim to foster and elevate the study of scientific disciplines at all educational levels. An EPA constitutes a formal pact between a defense laboratory and an educational institution to facilitate the transfer and improvement of technology applications and deliver technological assistance across the education spectrum, from pre-kindergarten onward. EPAs serve as the foundation for collaborative interactions between the government and the educational institutions, with a central focus on advancing and enriching science and engineering education. The law empowers the directors of each defense laboratory to establish EPAs with educational institutions in the US, with the express purpose of promoting and elevating the study of scientific disciplines at all educational tiers.

Cooperative Research and Development Agreement

The Stevenson-Wydler Technology Innovation Act of 1980 (PL 96-480) lay the foundation for cooperative research centers aimed at encouraging the utilization of technology advancements funded by the federal government – encompassing inventions, software, and training technologies. These were intended to benefit state and local governments and the private sector.

Subsequently, the National Technology Transfer and Advancement Act of 1995 (PL 104-113) (110 Stat. 775) amended the Stevenson-Wydler Act, enhancing the appeal of Cooperative Research and Development Agreements for both federal laboratories and private industry. This law offered US companies assurances regarding their intellectual property rights, thereby incentivizing the swift commercialization of inventions originating from CRADAs.

The governing statute for the Cooperative Research and Development Agreement is 15 USC 3710a. It authorizes federal labs to enter into agreements with other federal agencies, state/local government, industry, non-profits, and universities for licensing agreements for lab-developed inventions or intellectual property to commercialize products or processes originating in federal labs.

Other Transaction Agreements

The National Aeronautics and Space Act of 1958 (PL 85-568) established the National Aeronautics and Space Administration (NASA) and introduced other transaction (OT) agreements as a public procurement method to encourage innovation. Since 1958, other transactions (OT) have been used to encourage innovation and the development of new technology. Other transaction refers to any transaction other than a procurement contract, grant, or cooperative agreement. Although other transactions are not new, they are considered to be an alternative contracting vehicle for government agencies. The purpose of OTs is to help government agencies acquire leading-edge technology from private sector sources using a flexible, goal-oriented manner, which fosters new relationships through public-private partnerships. The three main benefits of OTs to the private sector are the decreased cost and time of the acquisition process, increased negotiating power for intellectual property rights, and more cooperation between the public and private sectors (Sargent Jr., Gallo, & Schwartz, 2018). This push for more cooperation between sectors and even between private sector firms positions the cooperative model (Bozeman, Crow, & Tucker, 1999) as a lens to examine innovation policies like those promoting alternative contracting methods.

Unlike other procurement contracts (FAR-based contracts), OTs are exempt from federal procurement laws and regulations (i.e., Competition in Contracting Act and Federal Acquisition Regulations (FAR)). Most federal government acquisition statutes and regulations do not apply to the OT authority, giving contracting officials more flexibility and freedom. For example, private sector companies are not subject to following the government accounting rules (also known as Cost and Accounting Standards (CAS) as prescribed by FAR part 30). In other words, the contract resembles a private sector (B2B) contract in which most terms and conditions can be negotiated.

Summary

An interest in policy and innovation led to increased inquiries and studies examining the relationship between policy and innovation. "From the 1970s onward, Douglas North, Robert Thomas, Nathan Rosenberg, and other economic historians argued that innovation was aided by specific government institutions and policies" (Taylor, 2016, p. 307). Policy implementation is the process of carrying out a government decision (Berman, 1978) by transforming a policy idea into an action intended to alleviate a social problem (Lester & Goggin, 1998). These actions may result in programs, procedures, or regulations (DeGroff & Cargo, 2009). Policymakers can use policy to promote innovation by using public procurement to stimulate innovative activity, especially among small businesses (Vonortas, Rouge, & Aridi, 2014, p. 16). Examining the outcomes of OTs can provide insight into how policy can promote the development of innovative technologies and products.

The connection between innovation and public-private partnerships is inherent (Roumboutsos & Saussier, 2014) because the relationship between innovation and public-private partnerships calls for the two sectors to "jointly develop products and services and share risks, costs, and resources which are connected with these products" (van Ham & Koppenjan, 2002, p. 598). As a result, this relationship should result in positive outcomes and some type of efficiency gains through the private sector's involvement in providing goods and services. Roumboutsos and Saussier (2014) make the point that public-private partnerships allow for sharing resources, knowledge, and risks to support innovation in ways traditional contracting activities cannot (Roumboutsos & Saussier, 2014). Many scholars have echoed the importance of the alignment of goals and values in public-private partnerships to help ensure that the public goals are met using this type of arrangement (Clark, Heilman, & Johnson, 1997; Kettl, 1993; Lombard & Morris, 2012; Savas & Savas, 2000).

OTs provide an alternative contracting method that is designed to promote more shared interest between the public and the private sectors through several innovation policies enacted by Congress. For example, Congressional actions that expanded definitions of nontraditional defense contractors to include small businesses and expanded the scope of prototype projects (PL 113-291, 2014) influence how DoD implements these policies through alternate contracting activities (i.e., OTs) to help promote the development of innovative technologies and products. The next chapter will take a deep dive into the Department of Defense authority as an example of leveraging public procurement and its various contracts and agreements to promote innovation.

Notes

1. Originally, this statute was 41 USC 505, but it was repealed by PL 97-258, §5(b), 13 September 1982, 96 Stat. 1083.

2. Code of Federal Regulations, Title 32, Subtitle A, Chapter I, Subchapter C, Part 21, Subpart F, Section 21.640, defining Cooperative Agreement. Available at: www.ecfr.gov/current/title-32/subtitle-A/chapter-I/subchapter-C/part-21/subpart-F
3. Formerly 10 USC 2371 with the same title.

References

Berman, P. (1978). The study of macro-and micro-implementation. *Public Policy*, *26*(2), 157–184.

Bozeman, B., Crow, M., & Tucker, C. (1999). *Federal laboratories and defense policy in the US national innovation system*. Paper presented at the Summer Conference on National Innovation Systems, Rebild.

Clark, C., Heilman, J. G., & Johnson, G. W. (1997). Privatization of wastewater treatment plants: Lessons from changing corporate strategies. *Public Works Management & Policy*, *2*(2), 140–147. doi:10.1177/1087724x9700200204

DeGroff, A., & Cargo, M. (2009). Policy implementation: Implications for evaluation. *New Directions for Evaluation*, 47–60. doi:10.1002/ev.313

Isett, K. R., Glied, S. A., Sparer, M. S., & Brown, L. D. (2013). When change becomes transformation: A case study of change management in Medicaid offices in New York City. *Public Management Review*, *15*(1), 1–17.

Jolly, J. A. (1980). The Stevenson-Wydler Technology Innovation Act of 1980 public law 96–480. *The Journal of Technology Transfer*, *5*, 69–80. doi:10.1007/BF02173394

Jong, J. P. J. D., & Hartog, D. N. D. (2007). How leaders influence employees' innovative behaviour. *European Journal of Innovation Management*, *10*(1), 41–64. doi:10.1108/14601060710720546

Kettl, D. F. (1993). *Sharing power public governance and private markets*. Washington, DC: The Brookings Institution.

Lester, J. P., & Goggin, M. L. (1998). Back to the future: The rediscovery of implementation studies. *Policy Currents*, *8*(3), 1–9.

Lombard, J. R., & Morris, J. C. (2012). Using privatization theory to analyze economic development projects. *Public Performance & Management Review*, *35*(4), 643–659. doi:10.2753/PMR1530-9576350404

Lutz Allen, S., Smith, J. E., & Da Silva, N. (2013). Leadership style in relation to organizational change and organizational creativity: Perceptions from nonprofit organizational members. *Nonprofit Management and Leadership*, *24*(1), 23–42. doi:10.1002/nml.21078

Paarlberg, L. E., & Lavigna, B. (2010). Transformational leadership and public service motivation: Driving individual and organizational performance. *Public Administration Review*, *70*(5), 710–718. doi:10.1111/j.1540-6210.2010.02199.x

Roumboutsos, A., & Saussier, S. (2014). Public-private partnerships and investments in innovation: The influence of the contractual arrangement. *Construction Management & Economics*, *32*(4), 349–361. doi:10.1080/01446193.2014.895849

Savas, E. S., & Savas, E. S. (2000). *Privatization and public-private partnerships*. Los Angeles, CA: SAGE Publications.

Sargent Jr., J. F., Gallo, M. E., & Schwartz, M. (2018). *The global research and development landscape and implications for the Department of Defense (R45403)*. Washington, DC: Congressional Research Service.

Singleton, R., & Straits, B. C. (1993). *Approaches to social research* (2nd ed.). New York: Oxford University Press.

Taylor, M. Z. (2016). *The politics of innovation: Why some countries are better than others at science and technology*. Oxford University Press.

van Ham, J. C., & Koppenjan, J. F. M. (2002). Port expansion and public-private partnership: The case of Rotterdam. *WIT Transactions on the Built Environment, 62*.

Vigoda-Gadot, E., & Beeri, I. (2012). Change-oriented organizational citizenship behavior in public administration: The power of leadership and the cost of organizational politics. *Journal of Public Administration Research and Theory: J-PART, 22*(3), 573–596.

Vonortas, N. S., Rouge, P. C., & Aridi, A. (2014). *Innovation policy: A practical introduction*. Springer.

Weber, M. (1947). *The theory of social and economic organization*. New York, NY: Simon and Schuster.

5 Other Transaction Authorities
A Department of Defense Case Study

Background

During the 1980s, the US experienced economic uncertainty, and scholars began to re-examine the private sector's role in innovation. The market failure paradigm began to lose its luster, and scholars needed a new way of assessing innovation policy. The challenger to the market failure paradigm was the cooperative model, which emphasized cooperation among sectors (Bozeman, Crow, & Tucker, 1999). This is important to note in the innovation policy timeline as the Department of Defense (DoD) assessed its challenges in the development of innovative technologies and products. Congress enacted Section 251 of Public Law No: 101-189, codified in 10 USC 2371 in 1989 (now codified in 10 USC 4021), which provided Defense Advanced Research Projects Agency (DARPA) the authority to conduct research and technology developments using cooperative agreements and other transactions. At the time, this authority was granted only for a two-year pilot program. In 1991, the National Defense Authorization Act of FY1992 extended 10 USC 4021 authority to other DoD agencies and made it permanent. This was the first major expansion of authority for OTs because it enabled other departments in the DoD to use OTs for their efforts.

The DoD's procurement contract method of promoting innovation is an excellent example of the market failure model and the mission model. The market failure paradigm has been the dominant model for innovation policy by assuming that if there is a need for innovation, then the private sector will "sense the need and respond in an economically efficient manner" (Bozeman et al., 1999, p. 6). Thus, for any non-mission-specific contracts, the DoD applies the market failure paradigm. The mission model becomes applicable for contracts that are mission-specific contracts (e.g., weapons systems). The mission paradigm assumes that the agency mission should derive innovation policy resulting in innovative products. These two models apply to the procurement contract method because they are not significantly different and are the most used innovation policy models.

The procurement contract method is a top-down regulated approach to procure goods and services from the private sector. A procurement contract

DOI: 10.4324/9781003398455-5

is a contract awarded according to the Federal Acquisition Regulation (FAR). Regulations governing the procurement contract method (i.e., the FAR, Cost Accounting Standards (CAS), and the Bayh-Dole Act) are adopted by any government agency that desires to do business with the private sector. This procurement contract approach should be familiar to scholars accustomed to the Mazmanian and Sabatier paradigm of the "single-authority top-down" methodology (Hjern & Hull, 1982). In this approach, "implementation is the carrying out of a basic policy decision, usually made in a statute" (Mazmanian & Sabatier, 1981, p. 540). Government support is proposed once a market failure is identified (Mazzucato, 2015). Later this government support is evaluated as an investment, and value is assessed using cost/benefit analysis. The six challenges identified in GAO-17-644 apply to the procurement contract method, another example of the top-down model.

As the cooperative model's popularity grew, the innovative policy promoting OTs as an alternative contract method evolved. Two years later, the National Defense Authorization Act of FY1994 (PL 103-160) expanded the OT authority for DARPA to include prototype projects related to weapons or weapons systems procured or developed for a DoD agency. This was significant because the extension of authority allowed OTs as a procurement method rather than a stimulant for research efforts. This legislation is the second major expansion of authority for OTs because it changed how the DoD was using OTs. Three years later, the National Defense Authorization Act of FY1997 (PL 104-201) extended Section 845 prototype authority across the entirety of the DoD.

Over the next two decades, the OT authority was expanded and restrained by enacting additional clarifications and reporting requirements for the transparency and accountability of DoD agencies. These reporting requirements included annual reports that would include a description of the transaction, the reason for not using a contract or grant to support the research, the amount of the payment, and other requirements.

The next major expansion of authority for OTs occurred nearly 20 years after the first major event, the National Defense Authorization Act of FY1994 (PL 103-160). In 2014, Carl Levin and Howard P. "Buck" McKeon National Defense Authorization Act for FY2015 (PL 113-291) expanded OT authority to include prototypes that were:

> directly related to enhancing the mission effectiveness of military personnel and supporting platforms, systems, components, or materials proposed to be acquired or developed by the Depart of Defense, or improvements of platforms, systems, components, or materials in use by the Armed Forces (p. 138).

Before this, OTs could only be used for weapons and weapons systems.

In 2015, the National Defense Authorization Act of FY2016 (PL 114-92) codified prototype OTs in 10 USC 4022, thereby rescinding the authority under

Section 845 and redefining and codifying non-traditional defense contractors in 10 USC 3014, and expanded follow-on production (10 USC 4022(f)). This legislation is the fourth major expansion of authority and is the focus of this study on how this innovation policy has promoted the development of innovative technologies and products. To highlight the importance of this study, it is important to review key previous research efforts over the past few decades, which is discussed in the following section.

The amount of the investment depends on available funding, and with the trends of the past several decades, the decline in funding for innovation has caused great concern. This is where the mission model and market model are challenged. Available financing, combined with the DoD's problem of not attracting innovative firms, can create a call for action. These challenges and the need for public and private sectors to work together illustrate the cooperative model's assumptions that the market is not always the most efficient way of promoting innovation.

For this reason, the cooperative model has gained interest, especially in the application in the commercialization of technology (Bozeman et al., 1999) and when addressing intellectual property and the government becoming a partner in developing the technology. These are not new topics and are found in numerous federal reports such as the reports mentioned earlier, the 2019 CRS report (R45521), and the 2017 GAO report (Sullivan, 2017).

Scholars such as Hjern and Hull (1982) who support the bottom-up approach to policy argue that examining only the perspective of "central" decision-makers neglects other actors who play roles in policy implementation (Sabatier, 1986). In a similar manner, one of the top-down models' critiques is that the model is weak when there is "no dominant policy, but rather a multitude of governmental directives and actors, none of the preeminent" (Sabatier, 1986, p. 30).

OTs provide a bottom-up emphasis, giving significant decision-making power to public administrators. These public administrators have considerable influence in shaping and enacting policy on the ground, especially when it lacks clear direction on its implementation (Hill, 2003). However, difficulties may occur in implementing policy in cases where "implementing agents know multiple ways to implement a policy and must choose among them" (Hill, 2003, p. 5). One significant reason is that OTs are not subject to the same rules and rely heavily on the contracting professionals' discretion to enforce and shape the policy governing these types of transactions. The authority provided by 10 USC 4022 gives contracting professionals decision-making power versus the regulation-driven procurement contract method.

OTs' appeal is that they are not procurement contracts but rather are an alternative contracting method consistently leveraged to help alleviate these types of concerns. In addition, OTs are designed to promote shared interests between the public and the private sectors through Congress's several innovation policies. For example, Congressional actions that expanded definitions of nontraditional defense contractors to include small businesses and increased the scope of prototype projects (PL 113-291, 2014) influence how the DoD implements these policies. In addition, these policies impact alternate contracting methods (i.e., OTs) to promote innovative technologies and products. This study will look specifically at the policy passed in 2015 in which Congress made OT authority permanent by codifying the law at 10 USC 4022 (PL 114-92, 2015). By using the cooperative model as a lens to examine how innovation policies such as 10 USC 4022 influence DoD alternate contracting activities, this study will also explore how this policy can promote innovative technologies and products using interrupted time series analysis.

Academics Tackling the Subject of Other Transactions

Most of the published OT literature reviewed was practitioner oriented (Bloch & McEwen, 2001; Dix, Lavallee, & Welch, 2003; Dunn, 2017; Kuyath, 1995; Vadiee & Garland, 2018), theory based (Schooner, 2002), or metrics based (Fike, 2009). The literature in peer-reviewed journals was focused on barriers for DoD's ability to keep pace with security needs in the current environment (Bell, 2014; Bonvillian & Van Atta, 2011; Michèle & Robert, 2016; Nunez, 2017; Peter, 2013; Steinberg, 2020; Steipp & Bezos, 2013), the legal and administrative systems that govern OTs (Gunasekara, 2010; Nathaniel, 2019; Nikole, 2019; Selinger, 2020; Victoria Dalcourt, 2019). Although these research subjects are important when discussing OTs, none of the OT literature reviewed attempts to systematically identify and discuss the impact of innovation policy on the DoD's use of OTs.

Steven Schooner published his contract law desiderata, arguing that three main policy goals of the US procurement system are transparency, procurement integrity, and competition (Schooner, 2002). Schooner notes that it is important to acknowledge the role of risk avoidance. Avoiding undue risk is the responsibility of the governing body. However, too much focus on risk avoidance can stifle creativity and innovation. These observations about the culture of DoD and its focus on avoiding undue risk provide some historical context of OTs as a potential solution to address certain institutional problems.

Several articles focused on the pros and cons of OTs (Bloch & McEwen, 2001; Dunn, 2017; Kuyath, 1995). Richard Kuyath (1995) is the earliest article developing a way to understand the pros and cons of OTs using data from program officials (Kuyath, 1995). David Bloch and James McEwen (2002) identified that OTs were created for three specific goals: enhancing military technological superiority, streamlining the procurement process, and integrating civilian and military technology industries (Bloch & McEwen, 2001). In his discussion in addressing the criticism of OTs, Richard Dunn (2017) provides case studies to help promote and encourage the DoD's use of OTs (Dunn, 2017). Gregory Fike's (2009) research attempted to find a reliable quantitative metric to assess DoD's OT effectiveness. He suggested several metrics to evaluate the success of an OT such as cost saving, time saved in negotiations, the procurement timeline, and participation of nontraditional contractors (Fike, 2009). This article was one of the few that attempted to provide actual metrics using a quantitative approach to measure the effectiveness of OTs.

Department of Defense Other Transaction Authorities

In November 1989, Congress enacted Section 251 of Public Law No: 101-189, codified in 10 USC 4022, giving DARPA the authority to conduct research and technology developments using cooperative agreements and other transactions. Congress later expanded this authority to the entire DoD to provide the department with the necessary flexibility to incorporate commercial industry standards and best practices into its award vehicle. Before 2015, the two types of DoD OTs were science and technology (S&T) authority and prototype authority (Section 845 PL 103-160). Today, there are three separate OT authorities that are used in different scenarios depending on the DoD's needs:

- 10 USC 4021 (Research OT) – used for basic, applied, and advanced research projects when procurement contract, grant, or cooperative agreement is not feasible or appropriate.
- 10 USC 4022 (Prototyping OT) – used for prototyping directly relevant to DoD mission requiring either one-third (1/3) cost share or significant nontraditional defense contractor (NDC) participation, and
- 10 USC 4022(f) (Production OT) – used for follow-on production contract or transactions authorized when a) competitive procedures are used for the selection of parties for participation in the transaction and b) the participants in the transaction successfully completed the prototype project provided by the transaction.

A best practice in assessing the use of OTs is to first understand the overarching purpose for their use. The OTAs were created to give federal agencies the flexibility necessary to adopt and incorporate practices that reflect commercial industry standards and best practices into its award instruments. When leveraged appropriately, OTs provide the government with access to state-of-the-art technology solutions from traditional and non-traditional defense contractors (NDCs) through a multitude of potential teaming arrangements tailored to the particular project and to the needs of the participants. A common misperception of the OT authorities is that their purpose is to engage small business, nontraditional contractors (NDCs), similar to a small business or socioeconomic set-aside program. While the use of OTs will often foster new relationships and practices involving traditional contractors and NDCs, especially those that may not be interested in entering into FAR-based contracts with the government, this is a benefit of OTs rather than the purpose of leveraging these authorities. For example, in a study in 2009, a recommendation was provided to find a more effective marker than the focus on nontraditional defense contractors such as performance characteristics, cost of the effort, and timeline (Fike, 2009).

The most important part of the procurement team's planning activities is defining the problem, area of need, or capability gap. This is critical in determining the correct acquisition pathway and the correct procurement vehicle to utilize in the acquisition strategy. When issuing a solicitation for a prototype OT, the government provides a problem statement, area of need or interest, or capability gap, and the industry submits a proposed solution. Depending on industry norms, the solutions proposed for a given problem may vary significantly in technical approach, schedule, and/or cost. The procurement team is responsible for understanding and clearly articulating to offerors the problem, area of need, or capability gap to allow for innovative trade space for a wide range of solutions. A best practice in this area is to allow for a wide area of solution tradespace by leveraging a statement of objectives, such as a description of the effort or even simply a problem statement.

Effective performance reporting addresses cost, schedule, and technical progress. It compares the work accomplished and actual cost to the work planned and the estimated cost and explains any variances. There is not a "one-size-fits-all" approach. There could be little, if any, performance reporting required if the agreement price is fixed, and fixed payable milestones provide financing. However, if this is not the case, performance reporting should be considered. For example, the awardee is responsible for managing and monitoring each project and all sub-awardees. The solicitation and resulting agreement should identify the frequency and type of performance reports necessary to support effective management. If an awardee is teaming with other sub-awardees (e.g., consortium, joint venture) for the project, the government team should consider if performance reporting on all sub-awardees would be appropriate.

The government team should also consider whether reports required of the OT awardee are important enough to warrant establishment of line items or separate payable milestones, or, if reporting requirements should be incorporated as a part of a larger line item or payable milestone. In either case, an appropriate amount should be withheld if a report is not delivered.

The 2016 Annual Industrial Capabilities Report[1] stated that Congress believed OTs could support the DoD's efforts to access new contractors of technological innovation, specifically with Silicon Valley startup and small commercial firms. The government's utilization of innovative companies in these alternative contracting vehicles is also critical because the government can create markets and engage private organizations that do not typically do business with the government, thus resulting in innovative technologies and products that are public goods.

Current Other Transaction Reporting Capabilities

Federal Procurement Data System – Next Generation (FPDS-NG) is the most commonly used source for the data because FPDS-NG reports contracts and agreements whose estimated value is $10,000 or more, including every modification to the respective contract, regardless of dollar value. It was used as a source outside of a common data collection source, because FPDS-NG provides procurement data to USASpending.gov, resulting in updated raw data on contract awards without including any grant or loan information that would be covered with USASpending.gov.

The sample of data used was raw data downloaded from FPDS-NG on 24 April 2021. The FPDS-NG data are reported by the DoD and are publicly available. The convenience of the data is a major advantage, and it helps mitigate risk associated with data collection and documentation. To process the data appropriately and to ensure that the method of analysis was applicable, the data were summarized using the Pivot Table feature in Microsoft Excel to run simple queries.

The first query examined the number of Contracting Agencies and their contract data. The downloaded data set included 8,587 records of total contract actions (i.e., contract awards and contract modification), totaling nearly $39 billion in obligated nominal dollars. Table 5.1 provides a breakdown of these numbers by contracting agency as well as the sum of obligated dollars for each agency.

Table 5.1 Contracting Agency Contracted Summary

Contracting Agency	Total # of OT Contract Actions	Total Dollars Obligated
Dept of the Army	4,409	$30,195,196,488.14
Defense Advanced Research Projects Agency (DARPA)	1,546	$2,160,016,064.11
Dept of the Air Force	1,008	$4,348,989,521.24
Dept of the Navy	807	$956,647,076.03
Washington Headquarters Services (WHS)	268	$604,004,943.80
US Special Operations Command (USSOCOM)	147	$77,607,065.12
Defense Contract Management Agency (DCMA)	146	−$61,536.52
Immediate Office of the Secretary of Defense	67	$76,362,833.79
Defense Information Systems Agency (DISA)	57	$71,474,930.20
Defense Threat Reduction Agency (DTRA)	45	$165,546,961.09
Missile Defense Agency (MDA)	32	$129,948,980.00
Defense Counterintelligence and Security Agency	21	$74,477,660.36
USTRANSCOM	14	$86,550,915.38
Defense Logistics Agency	13	$1,696,960.00
US Cyber Command	7	$3,605,364.65
Grand Total	**8,587**	**$38,952,064,227.39**

Out of the 8,587 records, only 1,852 were unique values identifying a DoD OT contract award. In other words, from May 2008 to 31 December 2020, DoD awarded 1,852 other transaction agreements and the remaining 6,735 were the respective modifications to those awarded OTs. This information was identified by applying the "Remove Duplicates" feature in Microsoft Excel within the Contract ID column. Using the same feature, information regarding the awarded vendors was captured. Over the 13-year period, the 1,852 unique contracts have been awarded to 1,013 unique vendors.[2]

Establishing the basis of analysis for the agencies is important to understand which agency within the DoD is most likely to use OT as a form of contracting. However, it does not provide data to tie back to the public law used to direct the decision-making and purpose of OTs. To gain an understanding of this, it is important to run queries regarding the data focusing on the vendors that are being awarded. As mentioned previously, 10 USC 4022 states that the DoD must ensure that prototype OTs meet at least one of the criteria; the significant contributing parties are either nontraditional defense contractor or small business. At least one-third of the total cost of the prototype projects is to be paid out of funds provided by parties to the transaction other than the federal government, or the senior procurement executive for the agency determines in writing that exceptional circumstances justify the use of an OT.

Table 5.2 shows the top ten awarded vendors based on Contract ID count include ten companies with at least 50 contract actions (this includes original award and modifications).

Table 5.2 Top Ten Awarded Vendors Based on Total Contract Actions

Vendor DUNS	Vendor Name	Total # of OT Contract Actions	Total Dollars Obligated
025172953	Advanced Technology International	1097	$19,815,056,065.44
827760138	SOSSEC, Inc.	758	$902,376,274.73
079799555	Consortium Management Group, Inc.	482	$1,747,809,840.61
180035768	National Center for Manufacturing Sciences, Inc.	320	$864,797,067.05
079639398	Defense Energy Center of Excellence	225	$380,159,662.87
079981146	Medical Technology Enterprise Consortium	200	$342,969,834.76
080331419	Defense Automotive Technologies Consortium	107	$266,733,884.17
794598573	Raytheon Company	82	$254,737,663.80
078824783	Pivotal Software, Inc.	81	$218,058,411.25
963411066	Consortium For Energy, Environment and Demilitarization	80	$215,067,959.07

Last, the items procured were analyzed by reviewing the Product Service Codes (PSCs). Table 5.3 shows that the top ten awarded PSCs based on contract actions (this includes original award and modifications). The benefit of this data is that it provides several years of longitudinal data from 2008 to 2020. It is important to note that in 2020, the government updated the PSC Manuals and retired hundreds of PSCs, many of which were used during 2008–2020.

This overview information describes the overall data and highlights interesting data points to assess whether innovation policies, specifically 10 USC 4022 (PL 114-92, 2015), influenced the DoD alternative contracting activities to promote the development of innovative technologies and products and fill the gap of academic and practitioner bodies of knowledge. To examine if time series analysis is appropriate, it is important to understand

Table 5.3 Top Ten Awarded PSCs Based on Total Contract Actions

PSC	PSC Description	Total # of OT Contract Actions	Total Dollars Obligated
AD92	Other Defense (Applied/Exploratory)/ R&D – Defense Other: Other (Applied Research/Exploratory Development)	1460	$1,400,334,710.20
AC54	Weapons (Engineering)/ R&D – Defense System: Weapons (Engineering Development)	1027	$10,730,577,143.83
AD94	Other Defense (Engineering)/ R&D – Defense Other: Other (Engineering Development)	720	$3,192,877,867.58
AD91	R&D – Defense Other: Other (Basic Research)	625	$586,674,308.73
AD93	Other Defense (Advanced)/ R&D – Defense Other: Other (Advanced Development)	360	$1,308,369,574.35
AZ14	R&D – Other Research and Development (Engineering Development)	289	$414,448,298.25
AZ11	R&D – Other Research and Development (Basic Research)	273	$211,734,373.69
AZ12	R&D – Other Research and Development (Applied Research/Exploratory Development)	217	$980,857,526.63
AZ13	R&D – Other Research and Development (Advanced Development)	209	$364,119,993.98
6910	Training Aids	200	$303,407,768.50

Table 5.4 DoD OT Contract Actions by Year

Year	Award	Mod	Grand Total
2008	2	1	3
2009	5	26	31
2010	17	58	75
2011	17	155	172
2012	23	127	150
2013	12	161	173
2014	16	213	229
2015	21	266	287
2016	35	331	366
2017	101	496	597
2018	265	732	997
2019	551	1,462	2,013
2020	787	2,707	3,494
Grand Total	1,852	6,735	8,587

how many contract actions (contract awards and modifications) have occurred over time. To assess that in a time series manner, another query was run using the Pivot Table feature in Microsoft Excel to examine if there was any type of linear trend among all the DoD OT contract awards over the course of 12 years. As a characteristic of time series data, the data must be in a sequence taken over equally spaced time. Table 5.4 provides a review of how the DoD awarded OT contract actions. Based on a visual review of the data, there is variation among the years. In 2017, there was a significant spike in the number of contract actions compared to the previous years. Comparing the number of contract actions totaling 366 in 2016 to 597 in 2017 resulting in a 63% increase in DoD OT contract actions.

When looking at the face value of these data, it is evident that their use is increasing; however, it may not necessarily explain why these changes are occurring. Unlike other US federal agencies, the DoD has updated their policy, guidance, and statute on the use of OTs frequently – making it difficult to infer causality – with the biggest change occurring in fiscal year 2018. As Congress stated, a method to measure effectiveness is required.

Measuring Innovation Policy Effectiveness

One method to measure this effectiveness is to see if certain changes create enough of a shift to impact the outcomes for an area of interest. This methodology is called a time series analysis and uses historical data over time (Remler, 2011), differentiating this analysis from cross-sectional studies. The overarching objective of time series analysis is to determine an appropriate model to describe a data pattern (Adhikari & Agrawal, 2013; Ramseyer, Kupper, Caspar, Znoj, & Tschacher, 2014). The model chosen in this study describes essential features of the time series pattern, explains how the past actions affect future actions, forecasts values, or identifies a control standard for quality.

Using interrupted time series research can provide insight to monitor policy outcomes by examining the effects of innovation policy enactment on prototype agreements. Using this quasi-experimental method, an interrupted time series analysis offers a practical way of evaluating the impact of already-implemented policies on outcomes. It is beneficial when a specific intervention has occurred at a particular time. The researcher's role is to assess whether the interruption had an impact on specific outcomes. Interrupted time series designs resemble one-group pretest-posttest design, except multiple observations before and after (Singleton & Straits, 1993, pp. 251–252). It is crucial to have the treatment applied systematically by using a naturally occurring intervention such as policy changes or other social changes. For example, "if a law had an impact, one would expect an 'interruption' or discontinuity in the time series . . . at the point where the law was introduced" (Singleton & Straits, 1993, p. 250). This study will use the 2015 enactment of 10 USC 4022 as the interruption. Information regarding the impacts of changes in policies is essential to policymakers, industry, and practitioners.

Measuring Impact on OT Awards

The National Defense Authorization Act for fiscal year 2016 was passed on 25 November 2015, nearly two months into the government fiscal year. Since the passing of the FY2016 NDAA, which codified 10 USC 4022, the number of prototype awards has increased each year. In 2017, the number of awards increased threefold totaling 101 awards compared to only 35 awards in 2016. In the following years, the number of awards doubled annually. The average annual growth rate (AAGR)[3] from 2008 to 2015 is 63%, compared to 114% for 2016–2020.

The Department of the Army had the most contract awards, totaling 743 new OT prototype awards from 2008 to 2020. However, since 2015, the Department of the Army increased the number of awards from 17 to 253 during 2016–2020, respectfully. DARPA was second with 372 awards from 2008 to 2020, increasing their awards from 13 to 136 during 2016–2020, respectfully. It is important to note that several new agencies started awarding prototype OT agreements after 2015, including Office of the Under Secretary of Defense (OUSDA), Defense Information Systems Agency (DISA), Defense Logistics Agency (DLA), Missile Defense Agency (MDA), US Cyber Command (USCYBERCOM), United States Transportation Command (USTRANSCOM), Defense Counterintelligence and Security Agency, and Defense Threat Reduction Agency (DTRA). These agencies are considered new since prior to 2015; these nine agencies had not previously awarded DoD prototype OT agreements, as shown in Table 5.5.

Table 5.5 DoD Prototype Awards by Agency (2008–2020)

Contracting Agency	'08–'15	'16	'17	'18	'19	'20
Dept Of the Army	50	17	73	145	205	253
Defense Advanced Research Projects Agency (DARPA)	55	13	21	70	77	136
Dept Of the Navy	2	0	0	20	112	156
Dept Of the Air Force	2	5	6	19	69	140
Washington Headquarters Services (WHS)	0	0	0	9	47	40
US Special Operations Command (USSOCOM)	4	0	1	1	21	29
Immediate Office of The Secretary of Defense	0	0	0	0	5	17
Defense Information Systems Agency (DISA)	0	0	0	0	4	8
Defense Logistics Agency (DLA)	0	0	0	0	2	5
Missile Defense Agency (MDA)	0	0	0	0	4	0
US Cyber Command	0	0	0	0	2	1
USTRANSCOM	0	0	0	1	1	1
Defense Counterintelligence and Security Agency	0	0	0	0	2	0
Defense Threat Reduction Agency (DTRA)	0	0	0	0	0	1
Grand Total (Annual)	113	35	101	265	551	787

OT awards changed dramatically in December 2015 when 10 USC 4022 was passed and the months following this policy change. In the 95 months prior to the full implementation, only 110 OTs were awarded. Most months saw no awards (38%) or a single award (32%). A few months saw two (17%) or even three (11%), but only three months saw four or more awards (range, 4–9). In contrast, in the 61 months that followed, there were 1,741 OT awards made, and rarely (<10%) were there no awards or only one award distributed (n = 4 and 1, respectively). In fact, following the intervention, a majority (60%) of months saw 10 or more awards, and about 30% of months saw 30 or more OT awards. The average number of OT awards prior to the implementation was 1.2 (s.d. = 1.4) and rose to 28.5 (s.d. = 34.1) following the passing of 10 USC 4022.

Given that the number of awards is highly skewed and that there is heterogeneity in the variance over time, two Autoregressive Integrated Moving Average (ARIMA) models were used – the first in its original metric and the second using the natural log of number of awards (see Table 5.6). Each model includes a constant, a linear trend (month), a dichotomous variable indicating pre (coded 0) and post (coded 1), and a linear component that begins with the change in policy – this is the critical variable in the model. Both models were used for all the hypotheses tested in Table 5.6. The impact of the policy change is not just statistically significant but has also a robust effect increasing the explained variance by 44% in the first and by 16% in the second logged model.

Table 5.6 Pre-Post Time Series ARIMA Model DoD Prototype OT Awards

	Original Metric			Natural Log of Total Award		
	Estimate	SE		Estimate	SE	
Constant	0.491	3.178		0.298	0.128	*
Linear Trend	0.014	0.057		0.006	0.002	**
Pre-Post	−14.937	5.065	**	−0.14	0.203	
Linear Post	1.33	0.126	***	0.055	0.005	***
R-Square	0.801			0.802		
Change the Linear Post	0.438			0.155		
Ljung-Box Q	111.61 w, 18 d.f., p < 0.001			36.85 w, 18 d.f. < 0.01		

Table 5.7 Pre-Post Time Series ARIMA Model DoD Prototype OT Awards ($)

	Original Metric			Natural Log of Total Award		
	Estimate	SE		Estimate	SE	
Constant	2,976,987	31,561,134		4.797	1.232	***
Linear Trend	(21,954)	570,920		0.066	0.022	**
Pre-Post	(70,985,783)	50,298,931	**	1.568	1.964	*
Linear Post	5,755,221	1,247,905	***	0.053	0.049	
R-Square	0.227			0.369		
Change the Linear Post	0.109			0.005		
Ljung-Box Q	5.716 w 18 d.f. p 0.997			16.697 w 18 d.f. p 0.544		

In ARIMA models presented in Table 5.6, the findings show statistical significance increase in the linear posttest, suggesting that that the passing of the policy in 2015 did increase the number of DoD prototype OT agreements. To understand this picture fully, an analysis of the total dollars obligated is also important.

OT awards changed drastically in December 2015 when 10 USC 4022 was passed, which is when the intervention began. In the 95 months prior to the full implementation, only 110 OTs were awarded with a total of $182.7 million obligated. Most months saw no awards (38%) or a single award (32%). A few months saw two (17%) or even three awards (11%), but only three months saw four or more awards (range, 4–9). In contrast, in the 61 months that followed, there were 1,741 OT with a total of $6.6 billion obligated. Looking further into the data, the average value of an award prior to the interruption was $1.6 million of obligated funds, and after the interruption, it increased to $56.6 million obligated funds, a 3,157% increase. The linear posttest (see Table 5.7) is statistically significant increasing the number of awards by just over $5.7 million each month, on average. The effect of the policy change is not just statistically significant but has also a substantial impact increasing the explained variance by 11% in the first and a small effect by 0.5% in the second logged model.

Similar to the number of awards, there is strong evidence to support that the passing of 10 USC 4022 in 2015 also impacted the dollars obligated. In the first ARIMA model presented in Table 5.7, the findings show a statistical significance increase; however, after the log transformation of the data, there was support for this hypothesis suggesting that after the passing of policy in 2015, there was an increase in the total dollars obligated of DoD prototype OT agreements.

Measuring Impact on New Entrants

This factor aims to test the companies awarded DoD OT agreements and the diversity of the organizations awarded those agreements. For purposes of this case study, new companies are defined as organizations that are recent entrants to DoD OT agreements. These companies include non-traditional defense contractors as defined in 10 USC 3014 and traditional defense contractors. Examining new companies awarded DoD OT agreements applies to innovation policy because it can provide insight into the diversity of organizations winning awards like DoD OTs and more. Using a similar method of analysis, the dollars obligated were examined for the first instance of obligation and removed from all duplicate records to capture only the total value of the first year the company won a contract. Examining FPDS-NG data, the total number of new companies awarded DoD prototype OTs increased significantly, from 28 to 385 throughout 2016–2020.

The diversity of companies changed drastically in December 2015 when 10 USC 4022 passed, which is when the intervention began. In the 95 months prior to the full implementation, only 78 new companies were awarded DoD prototype OTs. Most months saw no new companies (53%), or only one new company was awarded a DoD prototype OT award (49%). In contrast, in the 61 months that followed, there were 935 new companies that were awarded DoD prototype Ots, and rarely was there a month that had no new company awards (<1%). Furthermore, following the intervention, a majority (51%) of months saw 10 or more awards to new companies and about 30% of months saw 20 or more OT awards. The linear posttest (see Table 5.8) displays a statistically significant increase in the number of new companies being awarded DoD Prototype OT awards each month. The effect of the policy change is not just statistically significant but has a strong effect increasing the explained variance by 23% in the first and by 15% in the second logged model.

Examining the results presented in Table 5.8, both ARIMA models include findings showing a statistical significance increase. This hypothesis is intended to further dive into the results related to the impact of the policy on new companies receiving DoD OT awards. Based on the results presented here, there is support that following the codification of 10 USC 4022, there was an increase in the new companies receiving DoD prototype OT agreements.

Table 5.8 Pre-Post Time Series ARIMA Model DoD Prototype OT – New Companies

	Original Metric			Natural Log of Total Award		
	Estimate	SE		Estimate	SE	
Constant	0.464	1.66		0.276	0.122	**
Linear Trend	0.007	0.03		0.004	0.002	*
Pre-Post	−5.308	2.646	*	0.001	0.194	
Linear Post	0.62	0.066	***	0.046	0.005	***
R-Square	0.609			0.757		
Change the Linear Post	0.229			0.145		
Ljung-Box Q	100.923 w, 18 d.f., $p < 0.001$			53.129 w, 18 d.f., $p < 0.001$		

Measuring Impact on Types of New Products or Services Codes (PSCs)

The purpose of this factor is to examine the PSCs associated with DoD prototype OT agreements and the diversity of the product and service categories awarded by those agreements. PSCs are a measure of diversity and the impact of innovation policy, because the government categorizes all the products and services purchased by these codes. The government uses PSC codes to keep track of the types of products and services it procures for reporting purposes. An analysis of the PSCs from 2008 to 2020 provides a review of types of products and service categories procured with DoD prototype OT agreements. As DoD includes more PSCs, company diversity is clearly growing with the unique offerings coming into the agency. This is important to recognize for the encouragement of competition and overall improvement of product offerings in the marketplace. The top category based on the data is PSC AD92 – Other Defense (Applied/Exploratory)/R&D – Defense Other: Other (Applied Research/Exploratory Development). The activity for this PSC highly increased from 9 to 205 total awards from 2016 to 2020. The second-most awarded PSC is AD91 – R&D – Defense Other: Other (Basic Research), with 1–49 awards from 2017 to 2020.

OT awards changed dramatically in December 2015 when 10 USC 4022 was passed, which is when the intervention began. In the 95 months prior to the full implementation, only 21 new PSCs were awarded with a total of $59 million obligated. In contrast, in the 61 months that followed, 152 new PSCs were awarded with a total of $510 million obligated. Looking further into the data, the average value of an award prior to the interruption was $624,000 of obligated funds, and after the interruption, it increased to $8.3 million in obligated funds, a 1,230% increase. The effect of the policy change is statistically significant and has a substantial effect increasing the explained variance by 18% in the first and by 13% in the second logged model.

Table 5.9 Pre-Post Time Series ARIMA Model DoD Prototype OT – New PSC

	Original Metric		Natural Log of Total Award	
	Estimate	SE	Estimate	SE
Constant	0.272	0.264	0.173	0.085 *
Linear Trend	−0.001	0.005	−0.001	0.002
Pre-Post	−0.168	0.421	0.181	0.135
Linear Post	0.081	0.01 ***	0.024	0.003 ***
R-Square	0.559		0.608	
Change the Linear Post	0.177		0.133	
Ljung-Box Q	24.338 w 18 d.f. p 0.144		17.216 w 18 d.f. p 0.508	

In the ARIMA models presented in Table 5.9, the findings show a statistically significant increase in the diversity of product and service categories awarded DoD prototype OT agreements after the intervention in the linear posttest.

Other Notable Highlights

Unlike the procurement contract that is aligned with the mission model (Bozeman et al., 1999), OTs intend to provide benefits to the DoD such as attracting new companies, establish a network for resources to develop and/ or obtain innovative technologies, and provide an instrument for the DoD to influence technology and innovation. This is at the core of the cooperative model, which has been growing in popularity almost in parallel with DoD's effort to promote innovation through the expansion of OT authority.

The need to measure the impact of OT authority on public procurement has not only been the focus of academic and practitioner literature, but Congress also highlighted it as a recommendation in the 2019 CRS Report R45521. Several government reports have tried to examine the use of DoD OT agreements and their usage. These reports look at certain years using data from publicly available sources like this study, specifically FPDS-NG. Award data for DoD R&D in FPDS-NG was consistent throughout the years because FPDS-NG does not include DoD R&D data. After discussions with experts and a review of DoD public documents, the DoD implemented a deviation to the rule through C-Note: 20–03, which states all Research OTs, including modifications, are to be reported in Financial Assistance Award Data Collection System (FAADC), effective July 2019. Prior to that, the DoD recorded R&D OTs in the Defense Assistance Awards Data System (DAADS). Having various deviations and process adjustments throughout the years has an impact on the completeness of data available to the public and can create a validity issue if the Administrative Agreements Officer is not reporting accurately. It will be interesting to see if reporting requirements for R&D and prototype OT agreements will be adjusted again.

An important observation of the data is the fluctuation in the number of DoD prototype OT awards and the associated dollars obligated annually. The fluctuations may be a result of authority expansion since National Defense Authorization Act of FY2016 (PL 114-92), which permanently codifies OTs in 10 USC 4022, thereby rescinding the authority under Sec 845, redefines and codifies nontraditional defense contractors in 10 USC 2302(9), and expands follow-on production. Two years later, the National Defense Authorization Act of FY2018 (PL 115-91) added education and training requirements, increased approval thresholds, included language to clarify the approval levels applicable to OTs, and included express authority to allow for the award of prototype OTs in the SBIR program and non-profit research institutions. In addition, the FY2018 NDAA broadened the follow-on production language to include individual sub-awards under an OT consortium. Lastly, the John S. McCain National Defense Authorization Act for FY2019 (PL 15-232) removed USD (AT&L) as the highest-level approver and replaced it with USD (A&S) or USD (R&E) and clarified the application of follow-on production authority for projects carried out through Consortium Management Firms (CMFs). It is difficult to capture changes based on these policies, because it is not possible to differentiate prototype and production DoD OTs like you can between R&D and prototype OTs. Based on conversations with experts in the DoD audit agencies, the government does not have intentions to update the PIID nomenclature at this moment.

Another big focus of practitioner and academic literature is the government being able to attract new companies and nontraditional defense contractors to do business with the DoD. Per 10 USC 2302(9), nontraditional defense contractors are defined as an entity that has not worked or is not currently working with the government. This definition was updated the same year 10 USC 4022 was codified to provide more clarification regarding the definition of nontraditional defense contractors. One may defer that only small businesses would qualify under this definition, but in fact, many large businesses also qualify. In fact, the top contractors receiving OT prototype awards are consistent with the Top 100 Federal Contracts as reported by FPDS-NG. This guidance changed the definition, creating a hardship for contracting officers attempting to define further if the awarded company qualifies under the definition, especially with the addition of the language "and the regulations implementing such section, for at least a one-year period preceding the solicitation of sources by DoD for the procurement" (10 USC 3014). Now, contracting officers must contend with both large and small businesses, leading to more competition and time in the contracting process and again making it difficult to differentiate the status of the contractor winning DoD OT agreements.

Another key observation is the number of CMFs being awarded OT agreements. A CMF is defined as

> an association of two or more individuals, companies, or organizations participating in a common action or pooling resources to achieve a common

goal and can range from a handful to as many as 1,000 members. A consortium does not have to be a legal entity but must be legally bound through some form of teaming agreement or Articles of Collaboration.
(Department of Defense Inspector General, 2021, p. 3)

The privity of contract is with the prime entity doing business with the government; thus, reporting may not identify the performing party, only the managing party in the agreement. In other words, in a traditional principal-agent framework, the agent would typically perform the work under the contract arrangement. In a consortium, the agent is the contracting party, but they have an agreement in which a third party is performing the work. This has raised concerns in a recent US Department of Defense Inspector General's report titled "Audit of Other Transactions Awarded Through Consortiums" (dated 21 April 2021), Report No. DODIG-2021-077, and the results of this study showed that the top three awarded companies include consortia such as Advanced Technology International, SOSSEC, Inc., and Medical Technology Enterprise Consortium.

The 2021 DoD IG report used a sample of 13 base OT awards valued at $24.6 billion from 2017 to 2018 and found that these awards were not properly tracked, were not awarded in accordance with applicable laws and regulations, and were not consistent in negotiations of fees (Department of Defense Inspector General, 2021). The recommendations of the report provided that the DoD needs to develop policies for awarding and tracking OTs that are awarded to CMFs. These policies were intended to reinforce guidance, provide best practices, clarify current policies, establish controls for proper vetting, and develop procedures to review solicitations provided to CMF members. As stated previously, the Office of the Under Secretary of Defense for Acquisition and Sustainment is expected to release an updated OT guide for DoD in 2022, and it is suspected that guidance will be provided in the updated manual.

Lastly, the purpose of OTs is to bring innovative products and technology to the government for their use and commercialization. Like the concerns of regular policy changes and requirement updates through internal agency documentation, PSCs may be the most reliable measure across agencies on the product and service categorization. The most awarded PSCs have been retired as of 29 October 2020. Since 2015, over 839 PSCs have been retired, and 815 of those codes were retired on 29 October 2020, comprising 741 R&D PSCs and 27 IT PSCs, and the 17 remaining PSCs included maintenance, quality control, inspection, and leasing of equipment. In accordance with the Federal Procurement Data System Product and Service Codes (PSC) Manual dated October 2020, the 741 R&D PSCs that start with the letter "A" are being replaced by 155 new R&D PSCs that start with "A". Additional updates were made activating 23 new IT service PSCs and 17 new IT product PSCs. It is important to note that these PSC categories are used across all federal departments, not just the DoD.

Summary

The results of this case study show that innovation policy outcomes influence alternative contracting activities to promote the development of innovative technologies and products. The ARIMA models examined the role of the 2015 policy, 10 USC 4022, on DoD prototype OT awards, new companies receiving those awards, and the diversity of products and services associated with those agreements. There is strong evidence that after the passing of the policy in 2015, the number of DoD prototype OT agreements and the diversity of product and service categories awarded increased. There was also some strong evidence to support that after the passing of the policy in 2015, the dollars obligated to the DoD prototype OT agreements, and the diversity of new companies receiving these awards has increased. Through the lens of the cooperative model, it is clear that this policy was able to have an impact in one way or another. The policies themselves have room to grow through the early success of initial attempts. As some were proven more applicable than others, studies like these are necessary to enhance our knowledge base while encouraging change for more innovation. While continued research would allow for clearer results, the forward momentum and clear growth are important and should not be ignored.

From 2015 to 2020, the DoD significantly increased its use of prototype OTs in terms of number of DoD prototype OT awards and the amount of funds obligated for DoD prototype OTs. Nearly 70% of the dollars obligated were awarded to two traditional defense contractors and three consortia. The driving force for these changes is from the FY 2016 NDAA provision that expanded OT authority (10 USC 4022) to include follow-on production. Prior to 2015, DoD OT authority only covered R&D OT agreements, and all prototype authority was granted through Section 845 agreements. This meant that once a capability was developed that could move to full production, the government would have to use a traditional FAR-based contract. As a result of Congress codifying 10 USC 4022 in FY 2016 NDAA, follow-on production effort could be awarded without having to issue a traditional FAR-based contract.

Additional policy changes also affected the use of OTs in DoD. In the 2016 NDAA, Congress authorized the Small Business Innovation Research (SBIR) program to award prototype OT, provided clarification for OT approval levels within DoD and increased approval thresholds, and mandated additional training requirements. Following these changes, many DoD agencies such as the Army Contracting Command saw an increase in cross-service use of OT capabilities; however, there was no impact on overall OT adoption. Congress addressed this challenge in 2017 by including in the NDAA a mandate to increase the collection, storage, and reporting of OT usage data.

Other major contributing factors that may have impact on OT utilization are initiatives related to the US' near-peer adversaries, such as China and Russia. The DoD has been focused on addressing these threats specifically targeting acquisition speed and intellectual property (IP) considerations. According to one study, the threat of China's massive IP purchases is costing the US nearly $600 billion a year (Huang & Smith, 2019). In addition, since the COVID-19 pandemic, government agencies have been able to see the benefit of research initiated using OTs through the rapid development of the COVID-19 vaccine, which was a result of an OT (Soloway, Knudson, & Wroble, 2021). These environmental changes and policy changes external to the US example demonstrate the importance of government involvement in supporting innovative research and development of technology, hence, while supporting the cooperative model of interpretation. Due to regular policy updates, deviations, authority expansion, and limitations, it can be difficult to measure the effectiveness of a policy if policymakers do not look at data prior to revising current policy and/or introducing new policy. Significant steps are necessary to ensure that OT authorities achieve their purpose and impact is measured in accordance with the policy objectives.

The authority provided by 10 USC 4022 gives the contracting officers decision-making power outside of the procurement contract method, allowing more flexibility in their decision-making. This flexibility is highly dependent on the contracting officer. As pointed out by Montagnes and Wolton (2017), "[A] principal can choose a rule-based regulatory framework. However, unlike discretion, rules do not adapt to circumstances and are thus inefficient" (Montagnes & Wolton, 2017, p. 457). However, the argument in favor of the procurement contract method is that it provides a detailed process to ensure accountability and transparency. Policymakers require valid and relevant data to support their decision-making, a gap highlighted throughout 2019 CSR Report R45521. Interrupted time series analysis can examine the impact of a policy change after implementation, identify the changes the policy initiated, and illustrate any changes in the outcome over time. Providing information about the impact of policy can be essential in policy development. This includes bringing in data that is housed in other DoD databases such as DAADs and now FAADC. In addition, mapping the PSCs to their appropriate categories and possibly providing clarification or simplification to how products and services are categorized is a necessary effort. In public administration, the changes in innovation policy increased supplier diversity, meeting a continuous government acquisition goal to avoid company monopolization. In the procurement space, more dollars obligated, and efficient processes lead to sustained improvements, more outcomes, and technological advancements that can be built upon. The development of innovative technologies has a cyclical tie to our innovation policy, with one affecting the other, to promote new and positive outcomes within the procurement space. This implies moving toward efficient procurement timelines and defense technologies based on continuously updating innovation policies.

Notes

1. The 2016 Annual Industrial Capabilities report can be retrieved from https://www.businessdefense.gov/docs/resources/2016_AIC_RTC_06-27-17-Public_Release.pdf
2. One vendor, Advanced Technology International, had one Vendor DUNS and two Global DUNS. For purposes of this study, the Vendor DUNS was used as the identifier for the awarded vendors.
3. Average annual growth rate (AAGR) is the average annualized return of an investment, portfolio, asset, or cash flow over time. AAGR is calculated by taking the simple arithmetic mean of a series of returns.

References

Adhikari, R., & Agrawal, R. K. (2013). An introductory study on time series modeling and forecasting. arXiv:1302.6613.

Bell, R. L. (2014). Intellectual property in an emerging commercial spaceflight market: Taking advantage of other transaction authority to keep pace with changing commercial practices. *Public Contract Law Journal*, *43*(4), 715–735.

Bloch, D. S., & McEwen, J. G. (2001). Other transactions with Uncle Sam: A solution to the high-tech government contracting crisis. *Texas Intellectual Property Law Journal*, *10*, 195.

Bonvillian, W. B., & Van Atta, R. (2011). ARPA-E and DARPA: Applying the DARPA model to energy innovation. *The Journal of Technology Transfer*, *36*(5), 469–513. doi:10.1007/s10961-011-9223-x

Bozeman, B., Crow, M., & Tucker, C. (1999). *Federal laboratories and defense policy in the US national innovation system*. Paper presented at the Summer Conference on National Innovation Systems, Rebild.

Department of Defense Inspector General. (2021). *Audit of other transactions awarded through consortiums* (Report No. DEPARTMENT OF DEFENSE IG-2021-077). Department of Defense.

Dix, N. O., Lavallee, F. A., & Welch, K. C. (2003). Fear and loathing of federal contracting: Are commercial companies "really" afraid to do business with the federal government? Should they be? *Public Contract Law Journal*, 5–36.

Dunn, R. (2017). Appropriate contractual instruments for R&D. *The Government Contractor*, *59*(25), 1–4.

Fike, G. (2009). Measuring "other transaction" authority performance versus traditional contracting performance: A missing link to further acquisition reform. *The Army Lawyer*, 33–43.

Gunasekara, S. G. (2010). Other transaction authority: NASA's dynamic acquisition instrument for the commercialization of manned spaceflight or Cold War relic. *Public Contract Law Journal*, *40*, 893.

Hill, H. C. (2003). Understanding implementation: Street-level bureaucrats' resources for reform. *Journal of Public Administration Research and Theory: J-PART*, *13*(3), 265–282.

Hjern, B., & Hull, C. (1982). Implementation research as empirical constitutionalism. *European Journal of Political Research*, *10*(2), 105–115.

Huang, Y., & Smith, J. (2019, October 26). China's record on intellectual property rights is getting better and better. *Foreign Policy*. https://carnegieendowment.org/2019/10/16/china-s-record-on-intellectual-property-rights-is-getting-better-and-betterpub-80098

Kuyath, R. N. (1995, April 13). *The Department of Defense's other transactions authority under 10 US Code 2371*. Proceedings of the Panel Discussion of the Government Contracts Law Sections of the American Bar Association and the Federal Bar Association on "R&D Prototypes Using 'Other Transactions': Kiss the FAR Goodbye," pp. 521–577.

Mazmanian, D. A., & Sabatier, P. A. (1981). *Effective policy implementation*. Lanham, Mayland: Lexington Books.

Mazzucato, M. (2015). *The entrepreneurial state: Debunking public vs. private sector myths* (Vol. 1). Anthem Press.

Michèle, A. F., & Robert, P. L. (2016). Sustaining and enhancing the US Military's technology edge. *Strategic Studies Quarterly: SSQ, 10*(2), 3–13.

Montagnes, B. P., & Wolton, S. (2017). Rule versus discretion: Regulatory uncertainty, firm investment, and bureaucratic organization. *The Journal of Politics, 79*(2), 457–472. doi:10.1086/688079

Nathaniel, E. C. (2019). "Other transactions" are government contracts, and why it matters. *Public Contract Law Journal, 48*(3), 485–514.

Nikole, R. S. (2019). Jurisdiction over federal procurement disputes: The puzzle of other transaction agreements. *Public Contract Law Journal, 48*(3), 515–550.

Nunez, K. (2017). Negotiating in and around critical infrastructure vulnerabilities: Why the department of defense should use its other transaction authority in the new age of cyber attacks. *Public Contract Law Journal, 46*(3), 663–685.

Peter, J. M. (2013). Breaking the monopoly: The DoD's potential to reduce costs in its evolved expendable launch vehicle program. *Public Contract Law Journal, 43*(1), 87–104.

Ramseyer, F., Kupper, Z., Caspar, F., Znoj, H., & Tschacher, W. (2014). Time-Series Panel Analysis (TSPA): Multivariate modeling of temporal associations in psychotherapy process. *Journal of Consulting and Clinical Psychology, 82*(5), 828–838. doi:10.1037/a0037168

Remler, D. K. (2011). *Research methods in practice strategies for description and causation*. Thousand Oaks, CA: SAGE.

Sabatier, P. A. (1986). Top-down and bottom-up approaches to implementation research: A critical analysis and suggested synthesis. *Journal of Public Policy, 6*(1), 21–48.

Schooner, S. L. (2002). Desiderata: Objectives for a system of government contract law. *Public Procurement Law Review, 11*, 103.

Selinger, J. C. (2020). Closing the science and technology gap: Increasing non-federal participation in the intergovernmental personnel act mobility program through ethics reforms. *Public Contract Law Journal, 49*(3), 455–476.

Singleton, R., & Straits, B. C. (1993). *Approaches to social research* (2nd ed.). New York: Oxford University Press.

Soloway, S., Knudson, J., & Wroble, V. (2021). Other transactions authorities: After 60 years, hitting their stride or hitting the wall? *IBM Center for The Business of Government*. Retrieved from www.businessofgovernment.org/sites/default/files/Other%20Transactions%20Authorities.pdf

Steinberg, D. (2020). Leveraging the department of defense's other transaction authority to foster a twenty-first century acquisition ecosystem. *Public Contract Law Journal, 49*(3), 537–565.

Steipp, C. M., & Bezos, J. (2013). Funding cyberspace: The case for an Air Force venture capital initiative. *Air & Space Power Journal, 27*(4), 119.

Sullivan, M. J. (2017). *Military acquisitions: DoD is taking steps to address challenges faced by certain companies* (Report No. GAO-17-644). United States Government Accountability Office.

Vadiee, A., & Garland, T. (2018). The federal government's "other transaction" authority. *Thomson Reuters Briefing Papers–Second Series, 18*(5), 1–18.

Victoria Dalcourt, A. (2019). Innovation in government contracting: Increasing government reliance on other transaction agreements mandates a clear path for dispute resolution. *Public Contract Law Journal, 49*(1), 87–122.

6 Conclusion and Recommendations

Public Procurement and Innovation Policy Considerations

The government can directly support innovation through robust policies that leverage grants and contracts with agency-controlled funding. The federal government typically uses procurement contracts, which are contracts that are awarded according to the Federal Acquisition Regulation (FAR), to procure goods and services. However, innovation policy has also been known to promote alternative contracting vehicles, similar to the way the National Aeronautics and Space Administration (NASA), the Defense Advanced Research Projects Agency (DARPA), and the Department of Defense (DoD) have done in the past (e.g., Space Race, internet, Global Positioning System (GPS), and Siri). A great example of using alternative contracting vehicles occurred when President Eisenhower signed the National Aeronautics and Space Act of 1958 (PL 85-568) Section 203 (b)(5), granting NASA the authority to "enter into and perform such contracts, leases, cooperative agreements, or other transactions as may be necessary in the conduct of its work and on such terms as it may deem appropriate." This provision provided NASA with a flexible contract vehicle, known as other transactions (also called OTs), to procure innovative technology to combat the threat of the Soviet Union. This was the first mention of other transactions as an alternative contracting method for the federal government. Fast-forward to today, 12 government agencies, including the Department of Defense, use other transaction agreements for critical national security issues as Eisenhower did.

For example, other transactions aim to help government agencies acquire leading-edge technology from private sector sources using a flexible, goal-oriented manner to foster new relationships through public-private partnerships. The three main benefits of other transactions to the private sector are the decreased cost and time of the acquisition process, increased negotiating power for intellectual property rights, and more cooperation between the public and the private sectors. This push for more cooperation between sectors, and even among private sector firms, positions the cooperative model

DOI: 10.4324/9781003398455-6

Conclusion and Recommendations 67

mentioned in Chapter 2 as a lens to examine innovation policies promoting alternative contracting methods.

The realm of contract development and management has garnered increasing attention, driven by the imperative of budget reductions and the call for heightened operational efficiency. In the private sector, efficiency is sought through competition, while budget constraints and resource allocation shape the government's priorities. These contrasting dynamics play a pivotal role in comprehending public-private partnerships and the decision-making processes employed by each organization when considering outsourcing. Presently, the effectiveness measurement across all agencies revolves around tracking the number of contracts issued, their respective phases, the allocated financial resources, and other variables that may vary from one agency to another. These data must be reported to Congress via reporting mechanisms such as the Federal Procurement Data System – Next Generation (FPDS-NG) to ensure transparency. Subsequently, each federal agency or oversight body, such as the General Accountability Office (GAO), compiles this information into an annual public report. These reports assess various factors, the disparities among agencies, and their specific use cases. Some of these differences stem from agency enterprise-level variations, while others arise from distinct program management approaches. Policymakers are committed to evaluating these distinctions and promoting the adoption of best practices within agencies. The data extracted from these reports are invaluable for evaluating innovation development, utilizing these contractual avenues, and allocating federal funding.

Interest in the relationship between policy and innovation has grown, with scholars highlighting how government institutions and policies can support innovation. Policy implementation involves turning policy ideas into actions to address social problems, which can result in programs, procedures, or regulations. Policymakers can boost innovation, especially among small businesses, through public procurement. Observing the outcomes of Other Transaction Agreements (OTAs) provides insights into how policy can foster innovative technologies and products.

The connection between innovation and public-private partnerships is vital because these partnerships involve both sectors jointly developing products and services sharing risks, costs, and resources. This collaboration can lead to efficiency gains that traditional contracting may not achieve. Scholars emphasize the importance of aligning goals and values in these partnerships to ensure they meet public objectives.

OTs offer an alternative contracting method that promotes shared interests between the public and the private sectors, facilitated by Congressional actions such as expanding the definition of nontraditional defense contractors to include small businesses. These actions influence how the Department of Defense leverages public procurement to drive innovation, a topic explored in the Chapter 5 case study.

68 Conclusion and Recommendations

This case study underscores the significant impact of innovation policy on alternative contracting activities, particularly in promoting innovative technology and product development. The study employed ARIMA models to assess the influence of the 2015 policy (10 USC 4022) on Department of Defense (DoD) prototype Other Transaction (OT) awards. The results revealed a clear increase in the number of DoD prototype OT agreements and the diversity of product and service categories awarded after the policy's enactment. Moreover, the funds allocated to these agreements and the diversity of new companies receiving awards also exhibited growth.

The cooperative model of interpretation suggests that these policies can indeed make a difference, although there's room for further improvement as certain policies prove more effective than others. While ongoing research can provide more conclusive results, the evident positive momentum and growth should not be overlooked. From 2015 to 2020, the DoD significantly expanded its utilization of prototype OTs in terms of the number of awards and the funds allocated. This surge was largely driven by the FY 2016 NDAA provision that extended OT authority (10 USC 4022) to include follow-on production, eliminating the need for traditional contracts once a capability was developed. Additional policy changes, including Small Business Innovation Research (SBIR) program involvement, clarification of OT approval levels, and increased training mandates, also played a role. However, challenges remained in tracking and reporting OT usage. External factors, such as the threat posed by near-peer adversaries, further emphasized the importance of government involvement in supporting innovation and technology development. In addition, the COVID-19 pandemic showcased the benefits of research initiated through OTs, exemplifying the cooperative model's significance.

Overall, this case study highlights the need for continuous policy assessment and data-driven decision-making to ensure the effectiveness of innovation policies. The flexibility provided by 10 USC 4022 allows contracting officers to make decisions outside traditional procurement methods, but accountability and transparency remain vital. Valid and relevant data are essential for policymakers to support decision-making. Interrupted time series analysis proves useful in assessing policy impacts and guiding policy development. Moreover, supplier diversity and efficient processes resulting from innovation policy changes have a cyclical relationship with procurement advancements, promoting positive outcomes in the procurement space. This underscores the importance of continually updating innovation policies to drive technological advancements and efficient procurement practices.

This research is essential for several reasons. First, innovation is closely tied to economic growth related to a country's overall health. The US government uses various policy strategies to promote innovation. This research will seek to provide a new perspective on ways to encourage innovation through the lens of public administration and policy. The second reason is related to the current literature dealing with innovation. Academic literature

Conclusion and Recommendations 69

on innovation and innovation policy has predominantly originated from the economics perspective. However, R&D investments are cross-sector issues because public, private, and non-profit organizations contribute to funding and promoting innovation. The third reason is to encourage more research from diverse fields to promote policy innovation and assess how the different areas can collaborate to leverage knowledge and improve the US quality of life globally.

The challenge of public procurement and innovation is identifying those human needs and societal problems needing innovative solutions. Formulating a decision process on assumptions can be difficult for public organizations, especially when access to data is limited. One way to overcome this challenge is through interactive learning and "industry days" to promote discussion in a structured and focused group environment.

Events that promote cross-sector discussions, such as industry days, symposiums, and agency-run conferences, open the doors to communication in a collaborative manner, thereby leveraging the private market's resources and knowledge. This also minimizes the risk of information asymmetry in these contractual relationships. As mentioned in Chapter 3, information asymmetry occurs when one party has either more or better information than the other party in the relationship. Through creating a venue for open communications, public and private organizations can align goals and communicate expectations to formulate the need based on feedback and encourage transparency and accountability. These interactions should include representatives from relative fields, including potential users, researchers, policymakers, and other stakeholders in the value chain.

This book has merely scratched the surface of available points of interest. Delving deeper into the effects of policy changes on current contracting methods can expand our capabilities, technologies, and efficiency. As a recommendation, future studies must assess how much cooperation occurs between government and the private sector. Although the DoD has the authority, OTs' use is inconsistent across all agencies. Two agencies – the Army and DARPA – dominated OT awards from 2008 to 2020. Further review may indicate policy adoption readiness by agency and organizational inconsistencies. This is important because although the DoD OT authority has increased scrutiny, it has not been viably compared to other authorities to measure its effectiveness. Thus, additional research may show other gaps or successes in the policy that may indicate where the policy could be improved.

Academic research in public administration, business administration, economics, law, and other sciences can support such research by focusing on the importance of the topic. Many academic institutions do not include a focus on acquisitions, and many public administration journals do not highlight acquisition, public procurement, or innovation policy topics like their counterparts do in economics and law. Leveraging academia to assess policy further, contribute to policy writing, and measure policy effectiveness may improve adoption ratings among agencies and provide valuable metrics to promote more innovative policy. Using time series analyses provides an ability to forecast. Therefore, researchers can use time series data to predict future values of the dependent variables and help policymakers shape policy based on those predictions. This method allows policymakers to understand how policy changes affect contracting officers' decision-making. It can also provide insight into the discretion of agreements and contracting officers as more documentation clarifies the utilization and application of public procurement requirements.

This book is meant to serve as a primer to start the conversation about leveraging institutional knowledge on current policy implementation and using the proper metrics to assess policy effectiveness. Congress is interested in these measures. However, it is hard to measure the impact when those policies, regulations, and guidelines continue to change due to outside influence. An environment where our policymakers take the time to stop, assess, and create data-driven decisions can result in potentially decreased barriers of entry, less confusion on proper metrics, and a keen refocus of the nation on technological advancement versus tracking all compliance requirements.

Index

Note: Page numbers in **bold** indicate a table on the corresponding page.

acquisition: goal 62; instruments 37; pathway 48; process 10, 15–16, 26, 39, 72; reform 26; statues 17, 39
agreement, as contract 25–26, 27, 33–34, 40; *see also* cooperative agreement; other transaction; Partnership Intermediary Agreements
alternative contract method 13, 44
Alternative Contracting Pathways 32
Annual Industrial Capabilities Report to Congress (2016) 10, 16, 48
Armed Services Procurement Regulation 25
Army Contracting Command 16, 61
artificial intelligence (AI) 2–4
assistance agreement 35
automation 2–4
Autoregressive Integrated Moving Average (ARIMA): award hypothesis **55**; award models 54–56; award policy impact results **57–58**, 61, 73
average annual growth rate (AAGR) 53, 67n3

Bayh-Dole Act 13, 34, 44
Bloch, David 46
Broad Agency Announcement (BAA) 27

clean energy 3, 6
Code of Federal Regulations (CFR) 25, 27
competition: dependency on 5, 30; goal of 46, 57, 59, 72; limiting factors 24–25, 35; maximizing 21; prize 29
Competition in Contracting Act (CICA) 17
Congressional Research Services (CRS) 14, 45, 58
Consortium Management Firms (CMFs) 59–60
contract: actions 49, 51–52; awards 49–50, 52–53; development 21, 30, 72; law 15, 46; theory 24; *see also* procurement contract
cooperative agreement: distinguishing factor 36; guidelines 36, 71; research and development 12, 27, 37, 39, 43, 47
cooperative models 10–11
Cooperative Research and Development Agreement (CRADA) 33–34, 38–39
cost: analysis 13, 16, 27, 44; reducing 2, 5, 21, 39, 47–48, 50, 71–72
Cost Accounting Standards (CAS) 13, 16, 17, 39, 44
COVID-19 2, 62, 73
critical thinking 2, 30

Defense Advanced Research Projects Agency (DARPA): awards 53, **54**, 74; procurement process 71; research and development 12–13, 43–44, 47
Defense Assistance Awards Data System (DAADS) 58
Defense Contract Management Agency (DCMA) **49**
Defense Counterintelligence and Security Agency (DCSA) **49**, 53, **54**
Defense Federal Acquisition Regulation Supplement (DFARS) 15–16, 26, 30
Defense Information Systems Agency (DISA) **49**, 53, **54**
Defense Logistics Agency (DLA) **49**, 53, **54**
Defense Threat Reduction Agency (DTRA) **49**, 53, **54**
Department of Defense (DoD): award agreements 57–58; award benefits 58–60; award impact analysis 53, 56, **57**; award reporting 49–50, **52**, 52–53, **54–55**; cooperative agreements 36; grants 35–36; procurement process 43–48; prototype awards 53, **54**, 59; small business funding 28–29, 40
Department of Energy (DOE) 6, 29
Department of Health and Human Services (HHS) 29
Department of the Air Force **49**, **54**
Department of the Army **49**, 53, **54**
Department of the Navy **49**, **54**
Department of Transportation (DOT) 29
diversity 2, 56–58, 61–62, 73
Dunn, Richard 46

economic: growth 73; historians 5, 18, 40; objectives 21, 74; uncertainty 2, 11, 43
Edler, Jakob 18
Educational Partnership Agreement (EPA) 38

equal opportunity 2, 21
European Union (EU) 7
experimental 30, 53

Federal Acquisition Regulations (FAR): procurement (Part 12) 13, 15–16, 25–26, 44, 61, 71; research and development (Part 35) 27, 30
Federal Acquisition Streamlining Act of 1994 (FASA) 26
Federal Grant and Cooperative Agreement Act (FGCAA) of 1977 35–36
Federal Laboratory Consortium for Technology Transfer (FLC) 34
Federal Procurement Data System-Next Generation (FPDS-NG) 30, 49, 56, 58–59, 72
Federal Technology Transfer Act (FTTA) 34
Fike, Gregory 46
Financial Assistance Award Data Collection System (FAADC) 58, 62
fiscal year (FY) 10, 16, 27, 29, 52–53, 61, 73

General Services Acquisition Regulation Supplement (GSARS) 26
Government Accountability Office (GAO) 30, 45, 72; innovation policy report 18–644 10–11, 13, 16, 18, 44
grant (government funding) 14, 27, 35–36, 39, 44, 47, 49

Hjern, Benny 17, 45
Hull, Chris 17, 45

information asymmetry 6, 24–25, 74
innovation policy: challenges 16–17, 43; defining features 4–7, 18; effectiveness 52, 73–75; impact 46, 53, 56, 61; models 10–14, 43; policy outcomes 18, 40, 53, 61, 72; role of 1; trends 2–3

intellectual property rights 1, 10, 34, 39, 71
Internet 3
interrupted time series analysis 13, 46, 53, 62, 73

Kettl, Donald F. 25
Kuyath, Richard 46

labor 2, 25, 32
laboratory: defense 38; federal 12, 33–34, 38
Lane, Jan-Erik 24
leadership roles 5, 21, 32

market: arrangement 23; failure 11, 13, 43–44; model 11, 14, 18, 45; research 15, 26
Mazmanian, Daniel A. 13, 44
McCain, John S. 59
McEwen, James 46
methodology: innovation policies 19, 52; single-authority top-down 13, 44
Missile Defense Agency (MDA) **49**, 53, **54**
mission: contracts 13, 43; model 11, 13–14, 18, 43, 45, 58; statement 32
Montagnes, B. Pablo 62
Morris, John C. 24

National Aeronautics and Space Administration (NASA) 26, 39, 71
National Aeronautics Space Administration FAR Supplement (NFS) 26
National Defense Authorization Act (NDAA) 10, 12–13, 28, 53, 59, 61, 73
National Technology Transfer and Advancement Act of 1995 (NTTAA) 34–35, 39
NDAA *see* National Defense Authorization Act
Neill, Katharine A. 24
nontraditional defense contractor (NDC) 47

Office of the Secretary of Defense (OSD) **49**, **54**, 60
other transaction (OT): agreements 15–16, 53, 55–60, 73; authority 17–18, 30, 58, 71; contract actions 51–52; effectiveness 8, 44, 46–47, 52, 62; *see also* Other Transaction Authorities
Other Transaction Authorities (OTA) 7, 39, 72

Partnership Intermediary Agreements (PIA) 37–38
Patent and Trademark Law Amendments Act 34; *see also* Bayh-Dole Act
Peters, Heidi M. 2
policy implementation 17–18, 45, 75
policymakers: concerns 3–4, 62, 73–75; policy promotion 18, 39–40, 53, 62
principal-agent theory 24
private sector: behavior/influence 11–13; firms 17–18, 39, 71; growth 10; incentives 5, 24–27, 33, 39; role in innovation 43–45
prize competitions 29
problem-solving 2
procurement contract: flexibility 62; method promoting innovation 43–45; regulations 13, 15–17, 35
procurement policy 17
Product Services Codes (PSC) 51, 57, **58**, 60, 62
prototype project 13, 15, 40, 44, 46–47, 50
public procurement: challenges 32, 74–75; defined as 4–5; innovation 6–8, 18, 72; methods 39; strategic goals 21, 25, 30
public-private partnership (PPP): benefits 39–40, 71–72; definition of 22–24; innovation connection 18–19; leverage 25

regulation: awards 60
regulations: procurement 15–18, 21, 35, 39, 59; public/private policy 23, 72, 75; technology 3
request for proposal (RFP) 16, 21
research and development (R&D): defined as 4; funding 28–29, 33, 38; policy 10–11, 27
RFP *see* request for proposal
risk management 5, 21
robotic process automation (RPA) 4
Roumboutsos, Athena 19, 40

Sabatier, Paul A. 13, 44
Saussier, Stéphane 19, 40
Savas, Emanuel S. 22–23
SBIR *see* Small Business Innovation Research
Schooner, Steven 46
Schwartz, Moshe 2
science and technology (S&T) 47
science, technology, engineering, and math (STEM) 7
Secretary of Defense 49, **54**, 60
Simon, Herbert A. 25
Small Business Administration Policy Directive 28
Small Business Innovation Research (SBIR) 27–29, 35, 59, 61, 73
Small Business Technology Transfer (STTR) 28–29, 35
Socolar, Milton J. 36
solicitation 27, 48, 59–60
Stevenson-Wydler Technology Innovation Act 1980 33–34,

technological advancement 1–3, 7, 38, 62, 73, 75
Technology Investment Agreement (TIA) 37
technology transfer 12, 32–33, 38
timeliness 5, 21
traditional contracting 19, 40, 47, 56, 61, 72–73
transaction: cost 5, 12; *see also* other transaction

Under Secretary of Defense for Acquisition and Sustainment (OUSD (A&S)) 60
United States Transportation Command (USTRANSCOM) **49**, 53, **54**
US Congress: federal prize competition 29–30; public procurement/law 12, 15–16, 58, 61–62; research projects 37, 40, 43, 47–48; technology transfer laws 33–34
US National Security Strategy 7
US Special Operations Command (USSOCOM) **49**, **54**
US workforce 2, 7, 10

vendor awards 50

Washington Headquarters Services (WHS) **49**, **54**
Weber, Max 32
White House Office of Science and Technology Policy 6
Wolton, Stephane 62